99-HHE-1

EFFECTS OF WASTEWATER DISINFECTION ON HUMAN HEALTH

by:

Ernest R. Blatchley III and Woei-Long Gong
Purdue University

Joan B. Rose
Michigan State University

Debra E. Huffman
University of South Florida

Masahiro Otaki
Ochanomizu University

John T. Lisle
U.S. Geological Survey Center for Coastal and Watershed Studies

2005

WERF

The Water Environment Research Foundation, a not-for-profit organization, funds and manages water quality research for its subscribers through a diverse public-private partnership between municipal utilities, corporations, academia, industry, and the federal government. WERF subscribers include municipal and regional water and wastewater utilities, industrial corporations, environmental engineering firms, and others that share a commitment to cost-effective water quality solutions. WERF is dedicated to advancing science and technology addressing water quality issues as they impact water resources, the atmosphere, the lands, and quality of life.

For more information, contact:
Water Environment Research Foundation
635 Slaters Lane, Suite 300
Alexandria, VA 22314-1177
Tel: (703) 684-2470
Fax: (703) 299-0742
www.werf.org
werf@werf.org

This report was co-published by the following organizations. For non-subscriber sales information, contact:

IWA Publishing
Alliance House, 12 Caxton Street
London SW1H 0QS, United Kingdom
Tel: +44 (0) 20 7654 5500
Fax: +44 (0) 20 7654 5555
www.iwapublishing.com
publications@iwap.co.uk

Library of Congress Catalog Card Number: 2004104319

WERF ISBN: 978-1-893664-92-0 IWAP ISBN: 978-1-84339-673-4

Purdue University, Michigan State University, University of South Florida, U.S. Geological Survey Center for Coastal and Watershed Studies

ACKNOWLEDGMENTS

Report Preparation

Principal Investigator:
Ernest R. Blatchley III, *Purdue University*

Co-Principal Investigators
Joan B. Rose, *Michigan State University*

Debra E. Huffman, *University of South Florida*

John T. Lisle, *U.S. Geological Survey, Center for Coastal and Watershed Studies*

Project Team Members
Woei-Long Gong, *Purdue University*
Masahiro Otaki, *Ochanomizu University, Tokyo, Japan*

Project Subcommittee

Robert K. Bastian
Office of Wastewater Management, U.S. EPA

Chuck Gerba, Ph.D.
University of Arizona

Fred Bishop
National Risk Management Research Lab, U.S. EPA

Salvador J. Sedita, Ph.D.

Tom Marshall
City of Delaware, OH

Marylynn V. Yates, Ph.D.
University of California

William L. Cairns, Ph.D.
Trojan Technologies

Water Environment Research Foundation Staff

Director of Research: Daniel M. Woltering, Ph.D.
Project Manager: Lola Olabode, M.P.H.

ABSTRACT AND BENEFITS

Abstract:

The purpose of this research was to assess the effects of wastewater disinfection on human health. This assessment was conducted by examining two fundamental questions regarding the application of disinfection:

- Should municipal wastewater effluents be disinfected prior to discharge?
- When disinfection is necessary, how should it be accomplished?

Undisinfected effluent samples from a several municipal wastewater treatment facilities were collected to provide a broad spectrum of effluent quality, particularly as related to nitrogen compounds. Samples underwent bench-scale disinfection in compliance with relevant discharge regulations and so that disinfectant exposures could be accurately quantified. Disinfectants applied in this research included chlorine and ultraviolet (UV) radiation. Disinfected samples were subjected to a battery of microbiological assays to assess the immediate and long-term effects of disinfectant exposure on bacteria and viruses.

Taken together, and when considered in conjunction with previously published research, the results illustrate several important limitations of common disinfection processes as applied in the treatment of municipal wastewaters. It is not clear that disinfection processes, as commonly implemented, are effective for control of the risks of disease transmission, particularly as associated with viral pathogens. Microbial and chemical quality in receiving streams may not be substantially improved by the application of these methods; under some circumstances, disinfection may actually cause a decrease in the quality of effluent and, hence, receiving water. Decisions regarding the need for effluent disinfection must account for site-specific characteristics, but it is not clear that disinfection of municipal wastewater effluents is necessary or beneficial for all facilities. When direct human contact or ingestion of municipal wastewater effluents is likely (i.e., close to an outfall or in cases of discharge to bathing areas or food production areas), disinfection appears necessary. Under these circumstances, these experiments illustrate advantages of UV irradiation over chlorination in terms of microbial quality, as well as chemistry and toxicology. This advantage is more evident in effluents that contain appreciable quantities of ammonia-N or organic-N.

Benefits:

Water quality, in terms of chemical or microbiological constituents, can be compromised in receiving streams and reuse applications in the absence of or the improper application of municipal wastewater treatment. Among the various treatment operations that can be applied, disinfection can have an important effect on microbiological and chemical composition of effluent. However, there is considerable debate as to the overall benefits of disinfection, as applied to municipal wastewaters. To address these issues, a series of experiments was conducted to elucidate the effects of wastewater disinfection on effluent quality. The benefits of this research were as follows:

- Provides a detailed characterization of the effects of common disinfectants (chlorine and UV radiation) on common wastewater bacteria, in terms of initial response to disinfectant exposure, changes in the bacterial community post-exposure, and the nature and extent of bacterial physiological damage resulting from exposure to disinfectants.
- Provides a detailed characterization of viral (phage) responses to common disinfectants, and the nature of the viruses that retain viability.
- Examines the behavior of common disinfectants in wastewater effluents covering a broad range of treatment objectives, particularly as related to nitrogen removal.
- Characterizes the toxicological implications of chlorination/dechlorination in the presence of organic amines.
- Provides a quantitative assessment of risks of viral infection associated with the application or omission of disinfection as part of municipal wastewater treatment.
- Offers recommendations regarding the circumstances under which wastewater disinfection should be practiced.

Keywords: Bacteria, chlorine, disinfection, ultraviolet, virus, wastewater.

TABLE OF CONTENTS

LIST OF TABLES

LIST OF FIGURES

LIST OF ACRONYMS AND NOMENCLATURE

Abbreviations

ANOVA	analysis of variance
AOC	assimilable organic carbon
AOM	alogenic organic matter
APHA	American Public Health Association
ATP	adenosine 5'-triphosphate
BOD	biochemical oxygen demand
BRP	bacterial regrowth potential
CCL	contaminant candidate list
CI	confidence interval
Cl_2	chlorine
$Cl_{2(aq)}$	aqueous chlorine
CO_2	carbon dioxide
CoA	coenzyme A
COD	chemical oxygen demand
CPD	cyclobutyl pyrimidine dimer
DAPI	4',6-diamidino-2-phenylindole
DBP	disinfection by-product
$DiBAC_4$	bis-(1,3-dibutylbarbituric acid) trimethine oxanol
DNA	deoxyribonucleic acid
DO	dissolved oxygen
DPC	DNA protein crosslinks
DPD	(N,N-diethyl-*p*-phenylenediamine)
DSB	double-strand breaks
EI	electron ionization
FC	fecal coliform
GAC	granular activated carbon
GI	gastrointestinal
HAA	haloacetic acids
HOCl	hypochlorous acid
I_3^-	triiodide
KI	potassium iodide
LP	low pressure
MGD	million gallons per day
MIMS	membrane introduction mass spectrometry
MLSS	mass liquor suspended solids
MWRDGC	Metropolitan Water Reclamation District of Greater Chicago
N	nitrogen
NaOCl	sodium hypochlorite
NCl_3	trichloramine
NDMA	N-nitrisodimethylamine
NH_2Cl	monochloramine

$NHCl_2$	dichloramine
NH_3-N	ammonia-nitrogen
NRC	National Research Council
O_2	oxygen
OCl^-	hypochlorite ion
OUR	oxygen uptake rate
OUR test	O_2 utilization rate test
PBS	phosphate buffered saline
PRE	photoreactivating enzymes
PSS	point source summation
Rh123	Rhodamine 123
RNA	ribonucleic acid
SDWA	Safe Drinking Water Act
SSB	single-strand breaks
TCB	total culturable bacteria
TDC	total direct counts
THM	trihalomethanes
TiO_2	titanium dioxide
TSS	total suspended solids
TYL	tryptone lactose yeast
TYLD	tryptone lactose yeast with 1% deoxycholate
U.S. EPA	United States Environmental Protection Agency
UV	ultraviolet
v/v	volume to volume
w	with
WHO	World Health Organization
w/o	without
WRP	water reclamation plant
w/v	weight to volume
WWTP	wastewater treatment plant

Nomenclature

Δ	vertical distance from lamp axis to receptor site (cm)
ε_{254}	molar extinction coefficient of actinometer at 254 nm ($M^{-1}cm^{-1}$)
v_{254}	photochemical reaction rate at 254 nm (M/s)
φ_{254}	quantum yield of actinometer at 254 nm
ε_{352}	molar extinction coefficient of photochemical product at 352 nm with pathlength of 1 cm ($M^{-1}cm^{-1}$)
$\theta_{critical}$	maximum radial angle of exposed area (degree)
ρ_i	distance from i^{th} point source to receptor site (cm)
σ_q	absorbance coefficient for quartz (cm^{-1})
γ	degree of reduction
A	exposure area of the shallow dish reactor used in collimator experiment (cm^2)
ABS_1	observed absorbance of photochemical product at 352 nm under the

	collimator system
ABS_2	observed absorbance of photochemical product at 352 nm in the capillary flow reactor
C_1	concentration of photochemical product under the collimator system (M)
C_2	concentration of photochemical product in the capillary flow reactor (M)
d	diameter of opening of collimator (cm)
E_{abs}	radiation energy absorbed by actinometer (mJ)
EC_{50}	toxicant concentration estimated to cause a specified effect in 50% of test organisms (mg/L)
Gamma	values used account for the light loss in blank sample
H	collimator length (cm)
I (R,Z)	radiation intensity at position (R,Z) at wavelength λ ($\mu W/cm^2$)
$I'_{0,254}$	average radiation intensity available to actinometer at 254 nm ($\mu W/cm^2$)
$I_{a,254}$	rate of radiation absorbance at 254 nm ($\mu W/cm^2$)
$I_{reading}$	measured intensity at wavelength λ by radiometer ($\mu W/cm^2$)
I_{0c}, I_{tc}	light intensity of control after exposure time of 0 and t, respectively ($\mu W/cm^2$)
I_0, I_t	light intensity of sample after exposure time of 0 and t, respectively ($\mu W/cm^2$)
L	lamp arc length (cm)
l	radiation pathlength (cm)
n	number of point sources
P	lamp output power at wavelength λ (W)
P'	the ratio of equivalent dose of biodosimetry test to calculated dose by PSS model
Q	sample flow rate (mL/min)
R	radial distance between quartz tube and UV lamp (cm)
S	radial distance between lamp and side wall of capillary flow reactor (cm)
S_a	concentration of actinometer (M)
t	thickness of the side wall (cm)
t_{exp}	exposure time (sec)
t_q	wall thickness of quartz tube (cm)
V_1	volume of shallow dish reactor used in collimator experiment (cm^3)
V_2	exposure volume in the capillary flow reactor (cm^3)
$Y_{X/S}$	growth yield factor (g/g)
z	longitudinal distance along the quartz tube (cm)
z_f	final position of point source which contributes to $I_{reading}$ (cm)
z_i	initial position of point source which contributes to $I_{reading}$ (cm)

EXECUTIVE SUMMARY

Wastewater disinfection is generally viewed as a method of protecting human populations from exposure to pathogenic waterborne microorganisms. When properly applied, disinfection can bring about an immediate, substantial reduction in the concentration of viable waterborne pathogenic microorganisms. Disinfection processes have been viewed as an important component of disease prevention.

Disinfection operations have the potential to be an important barrier against human exposure to microbial pathogens, at least for a short period following disinfectant exposure. However, questions remain about to the overall benefits of wastewater disinfection and the need for its application as a default process in municipal wastewater treatment facilities.

Important limitations of wastewater disinfection have been identified, including the existence of microbial pathogens that are resistant to common disinfectants, the presence of common wastewater constituents that can adversely affect disinfection process performance, microbial repair and recovery following disinfectant application, and adverse effects of disinfectants on wastewater effluent chemistry and toxicology.

Other anecdotal considerations are relevant when considering the need for disinfection. For example, wastewater disinfection is infrequently applied in many developed countries (e.g., most of Western Europe), yet the incidence of disease associated with waterborne pathogens in these countries is not substantially higher than in North America, where wastewater disinfection is common. Even in North America, wastewater disinfection is not always practiced. For example, in the temperate states, seasonal application of municipal wastewater disinfection is common. This practice of operating disinfection systems during "warm weather" months is justified on the basis of the likelihood and frequency of direct human contact with receiving water. Yet seasonal disinfection is applied in many facilities that discharge effluent upstream of potable water intakes. There is no evidence that these discharges result in any change in the incidence of communicable disease among the downstream water users.

However, even in warm-weather months, there is some debate about the benefits of wastewater disinfection, in terms of human health and the environment. For example, the Metropolitan Water Reclamation District of Greater Chicago (MWRDGC) operates municipal wastewater treatment facilities that at times comprise up to 90% of the total flow in receiving streams. Yet, disinfection operations have been discontinued at MWRDGC facilities. The decision to cease chlorine disinfection operations at MWRDGC's facilities was based largely on the results of extensive investigations of the effects of disinfection on microbiology, chemistry, and wildlife in MWRDGC's receiving water. These studies demonstrated that water quality in the receiving stream, as measured by concentrations of various bacterial, viral, and chemical constituents, was not significantly affected by the application of chlorine, except near the point of effluent discharge. Moreover, MWRDGC demonstrated that cessation of disinfection actually led to important improvements in water quality, as defined by fish and wildlife populations (e.g., population size and species diversity). Based on this information, MWRDGC successfully petitioned the Illinois Environmental Protection Agency to allow cessation of its chlorine-based disinfection operations. Subsequent monitoring of effluents and in receiving streams indicates that water quality has not deteriorated; some water quality characteristics have actually improved.

Beyond these concerns, it is important to consider variations in disinfectant processes among wastewater treatment facilities. The overall behavior of a disinfection system will be affected by (undisinfected) effluent composition, the type of disinfectant applied, the design of the disinfection system, and the operating conditions. For example, the presence or absence of nitrogenous compounds (organic or inorganic) can have a profound effect on chlorine-based systems. Chlorinated forms of these compounds are generally less effective disinfectants than free chlorine. Moreover, inorganic and organic nitrogen compounds represent important precursors to disinfection by-product (DBP) formation. Nitrogenous compounds can also have an adverse effect on ultraviolet (UV) disinfection systems as UV-absorbing compounds.

The responses of microorganisms can vary dramatically among various forms of disinfection. Hence, disinfection operations can yield vastly different changes in effluent microbial and chemical quality. Therefore, when considering disinfectant application in a municipal wastewater treatment setting, it is not clear that disinfection processes are always beneficial.

The purpose of this investigation was to address two central questions:

- Should municipal wastewater effluents be disinfected prior to discharge?
- When disinfection is necessary, how should it be accomplished?

To address these questions, a series of experiments was conducted to investigate important microbiological responses to common disinfectants (chlorine and UV). In addition, experiments were conducted to elucidate relevant chemical and toxicological responses of chlorine-based disinfection systems. Data regarding viral responses to disinfectants were used as input parameters in a risk assessment model.

Testing of microbiological responses to disinfectants was conducted on effluent samples collected from various treatment facilities. Samples were collected from five municipal wastewater treatment facilities that spanned a range of treatment objectives and effluent qualities. A particular focus in the selection of these facilities was the degree to which biochemical treatment accomplished removal of (inorganic and organic) nitrogenous compounds. Facilities from which samples were collected spanned the spectrum of nitrogen removal processes, from non-nitrified effluents, to effluents that had been subjected to biological nitrification, to an effluent that was subjected to nitrification/denitrification. Non-disinfected effluent samples were collected from each of the facilities and subjected to bench-scale disinfection under conditions that allowed compliance with relevant discharge permit regulations and under conditions where disinfectant exposure could be accurately quantified. Disinfectants applied in these experiments included chlorine and UV.

Samples from these bench-scale disinfection operations were then subjected to an array of tests to characterize microbial responses to disinfectant applications. Bacterial populations and total (aerobic) microbiological activity were monitored over the course of six days in an incubated environment. Specific physiological responses of several bacteria were also monitored. The responses of indigenous phage were characterized and quantified.

Collectively, the results of these experiments revealed a number of interesting features of the responses of bacteria and viruses to common wastewater disinfectants. For non-nitrified wastewater effluents, disinfection processes (chlorination/dechlorination or UV) significantly reduced long-term oxygen consumption, probably as a result of disinfectant-induced stress on nitrifying bacteria. This effect was most pronounced for UV irradiated wastewater effluent samples. For wastewater effluents containing low ammonium concentration, disinfection processes did not significantly reduce long-term oxygen consumption. Of course, disinfection led to a decrease in viable fecal coliform concentration immediately following disinfectant exposure. However, disinfected samples also generally had lower viable fecal coliform concentrations than non-disinfected samples after five days of incubation; chlorination and UV irradiation were generally statistically indistinguishable in terms of viable fecal coliform concentration after five days of incubation. The behavior of the total bacterial community was substantially different than the behavior of fecal coliforms. Chlorinated samples generally had the highest total bacteria concentration among all treatments after five days of incubation. Although UV irradiated samples generally had lower total bacterial concentrations after incubation than the chlorinated samples, some of the UV irradiated samples showed higher total bacterial concentrations than controls (non-disinfected samples) after incubation. Results indicate that disinfection processes (chlorination/dechlorination or UV) can effectively reduce fecal coliform concentration; however, they have little or an adverse effect in terms of total bacteria. Generally speaking, when fecal coliform bacteria were used as the indicator, disinfected samples became more "unstable" or had higher regrowth potential than undisinfected samples when compared with initial values (t=0 hour). However, disinfected samples generally had lower absolute fecal coliform concentrations than undisinfected samples.

Several non-culture-based assays were used to develop more detailed information regarding the responses of bacteria to these disinfectants. These experiments revealed that common indicator bacteria (e.g., fecal coliforms and enterococci) generally represent only a small fraction of the total bacterial population. Moreover, the responses of these indicator species to disinfectants were not representative of the general bacterial population. The use of total culturable bacteria (TCB) is suggested to represent a more sensitive and appropriate microbiological assay for assessment of disinfection efficacy and microbial inactivation rates. However, the assay requires 14 days of incubation to provide reliable results.

It is believed that viruses in municipal wastewater represent a greater risk to human health than bacteria. As an alternative to conducting measurements of human viruses in wastewater samples, assays for (viable) indigenous phage were performed. In general, UV irradiation was more effective in controlling phage than chlorination/dechlorination. This trend was evident in effluent samples from all facilities, but was most pronounced in samples where ammonia-N removal (i.e., nitrification) had not been practiced. In particular, these results indicated that under the conditions commonly used for wastewater disinfection, UV irradiation was found to be more effective at controlling viruses (or at least phage) than chlorination in terms of total phage concentration and diversity. These data, which were based on analysis of naturally-occurring phage in real effluent samples, also support the conclusions that have been made by others pertaining to the antiviral characteristics of chlorine; namely, (inorganic) combined chlorine is less effective than free chlorine for control of viruses, and either form of chlorine (free or combined) is generally less effective than UV irradiation for virus (phage) inactivation.

Given the virtually unlimited spectrum of chemicals that could be present in wastewater effluents, it is not possible or practical to attempt a complete assessment of the chemistry or relevant toxicology of all wastewater constituents. As a surrogate approach, experiments were conducted to characterize the toxicological implications of several organic amines. These compounds were chosen because amines represent an important fraction of organic nitrogen and because relatively little is known regarding the chemistry of organic amines in disinfection systems, particularly relative to their interactions with chlorine.

Four aliphatic amines were chosen for investigation. Chlorination of these compounds appeared to proceed largely via substitution pathways, whereby +1-valent chlorine was N-substituted onto the amine. The chemistry of these reactions appeared to be similar to the reactions that are responsible for formation of inorganic chloramines. Toxicological responses of the parent compounds and their chlorinated forms were measured using two different toxicity assays: an acute Microtox assay, and a respirometric assay. In both assays, the chlorinated forms of these compounds were shown to be roughly one order of magnitude more toxic than their parent compounds.

The results of this research allowed characterization of some of the important chemical and microbiological changes that can be expected to result from the application of conventional disinfectants at municipal wastewater treatment facilities. Of course, the overall objective of disinfection is to reduce risks to human health (and presumably the environment). A risk model was developed to quantify the effects of disinfection (by chlorination/dechlorination or UV irradiation) on the risks of disease associated with microbial pathogens. This modeling effort demonstrated that the application of these conventional disinfectants, under conditions that allow compliance with existing effluent discharge regulations, allows a quantifiable decrease in the risks of disease transmission associated with viruses; however, the risk reductions accomplished by disinfection were modest in all cases. It was not possible within the scope of this project to conduct analogous risk assessments based on exposure to other microbial pathogens or chemicals (e.g., DBPs).

Collectively, the findings of this research illustrate several important limitations of wastewater disinfection practices. When considered in the context of the central questions of this research, these data suggest the following:

- Decisions regarding the need for effluent disinfection must be made on a site-specific basis; it is not clear that wastewater disinfection should be practiced in all cases. Control of microorganisms appears to be modest, and in the case of bacteria, is apparently confined to a relatively short period following disinfectant exposure. Disinfection is warranted in situations where direct human contact in the immediate vicinity of an outfall is possible. Effluent discharge to areas involved in food production are likely to benefit from wastewater disinfection. Important examples of these situations include reuse applications and effluent discharge to shellfish breeding grounds. Beyond these circumstances, the need for application of disinfection as a default process it is not as clear, at least not based on the disinfection alternatives that are most commonly applied today (chlorination and UV irradiation) and the methods used to design and monitor these systems (i.e., control of bacterial indicators and residual chlorine concentration). Disinfection is warranted in situations where its application leads to a reduction in the risk of disease transmission. Given the limitations of these practices, as illustrated by post-disinfection regrowth of bacteria, relatively poor virucidal behavior, and (in the case of chlorination) generation of persistent DBPs, it is not clear that wastewater disinfection always yields improved effluent or receiving water quality.

- In circumstances where disinfection is deemed necessary, UV irradiation appears to represent a better choice than chlorination/dechlorination. UV irradiation is more effective for control of viral pathogens than chlorination/dechlorination. Moreover, long-term bacterial control appears to be more effectively accomplished with UV irradiation than with chlorine-based practices. From a chemical and toxicological perspective, UV irradiation appears to present considerable advantages over chlorine. However, the design and implementation of disinfection at a wastewater treatment facility must take into site-specific issues and constraints that may influence the choice of disinfection technology.

CHAPTER 1.0

INTRODUCTION

1.1 Background

Wastewater disinfection has been practiced in the United States for approximately 100 years with the goal of protecting human populations from exposure to pathogenic waterborne microorganisms. Although bacteria were discovered in the late 1600s, it was not until approximately 1880 that a link was established between specific bacterial species and specific diseases. Non-disinfected wastewater effluents represent a potentially important source of pathogenic microorganisms in the environment and a possible vector for transmission of disease among human populations.

Although bacterial pathogens are present in wastewater effluents, it is generally believed that enteric viruses represent the greatest risk to human health of all waterborne pathogens. The (oo)cysts of waterborne protozoa can also represent a substantial risk to humans. The risks of disease associated with waterborne pathogens are generally acute in nature.

In considering the effects of disinfection on human health, it is also important to consider the impact of disinfection processes on water chemistry. All chemical and photochemical disinfection processes generate disinfection by-products (DBPs). Although the chemistry of DBP formation demonstrates site-specificity, chlorine-based disinfection processes are generally believed to generate DBPs that have more serious human health and environmental toxicology implications than those of other disinfection processes. Many chlorinated DBPs are persistent in the environment and are likely to be incorporated into downstream drinking water systems. Because municipal wastewater effluent is part of the same water cycle as potable water, the chemical and microbiological composition of treated wastewater effluent is relevant to drinking water systems. In contrast with the risks associated with microbial pathogens, human health risks associated with chemical constituents (including DBPs) tend to be chronic. Analyses of the effectiveness of disinfection processes must consider the balance between microbiological and chemical risks.

The risk of disease transmission to human populations is associated with human contact with pathogen- or DBP-containing water. Direct human contact with municipal wastewater effluent can occur through ingestion, swimming, direct or indirect contact from water reuse applications, or ingestion of seafood or freshwater fish. Beyond direct contact, it is important to consider that municipal wastewater effluent becomes part of the hydrologic cycle. As such, wastewater effluent represents a potentially important source of biological and chemical constituents in water supplies. It is important to keep in mind that "we all live downstream."

Among disinfection processes, chlorination has been used most extensively. The earliest recorded practice of large-scale wastewater chlorination was in 1854 when the Royal Sewage Commission used chloride of lime to deodorize London sewage (Hofman and Frankland, 1859).

Chlorine exists in different forms in water, and is classified as free chlorine and combined chlorine. Free chlorine includes hypochlorous acid (HOCl), hypochlorite ion (OCl^-), and aqueous chlorine ($Cl_{2(aq)}$); however, $Cl_{2(aq)}$ is generally present in insignificant concentrations in wastewater matrixes. If nitrification is not practiced at a wastewater treatment facility, appreciable quantities of ammonia nitrogen (N) and organic-N are likely to be present in the effluent. Effluent ammonia-N concentrations in non-nitrified effluents are typically 10 mg/L to 40 mg/L (as N). Free chlorine may react with inorganic- and organic-N compounds to form inorganic monochloramine (NH_2Cl), dichloramine ($NHCl_2$), trichloramine (NCl_3), as well as a number of organic chloramines. Aqueous combined chlorine is often quantified by titration and concentrations are usually expressed as the equivalent chlorine concentration (as Cl_2). However, the titrimetric endpoints that are normally ascribed to inorganic chloramines are susceptible to interference from a number of other compounds that are likely to be present in water, most notably organic chloramines (e.g., Shang and Blatchley, 1999).

Dechlorination is necessary to reduce effluent toxicity, since it is known that free chlorine and chloramines can cause acute toxicity effects in receiving waters. Dechlorination is generally accomplished by adding dechlorination reagents, usually in the form of reduced sulfur (e.g., sodium bisulfite), followed by discharge to the environment. Studies have indicated that some organic chloramines are recalcitrant to S(IV)-based dechlorination and may express toxicity in dechlorinated wastewater effluent. This suggests that organic chloramines might pose toxicity (Helz and Nweke, 1995). However, relatively little quantitative information is available to directly link observed toxicological responses to specific organic chloramines.

Free chlorine, with its strong oxidation power, will react not only with microorganisms, but also with natural organic matter (NOM) to form a variety of DBPs, such as trihalomethanes (THMs), haloacetic acids (HAAs), and haloacetonitriles (HANs) (Ueno et al., 1996). Inorganic chloramines, which are formed during chlorination of inorganic-N compounds (ammonia-N), express less oxidation power than free chlorine. However, previous research has shown that chlorine transfer between inorganic chloramines and some organic-N compounds can be faster than free chlorine (Scully and Hartman, 1996). Further, inorganic chloramines can react with organic chloramines to form cyanogen chloride (Krasner et al., 1989).

Although chlorination can usually achieve satisfactory inactivation of indicator bacteria, formation of many undesired (toxic or carcinogenic) by-products has led to a re-evaluation of its risk to human health. Further, recent studies have indicated that some viruses and protozoa, such as *Cryptosporidium parvum*, are highly resistant to conventional chlorine-based disinfectants (Roefer et al., 1996; Pett, 1993). These known limitations of chlorine-based disinfection have motivated the consideration of alternative disinfection processes that are more “friendly” to human health and the environment. The most promising of these alternatives may be ultraviolet (UV) irradiation, because it can inactivate pathogenic microorganisms effectively and is generally not linked to the production of effluent toxicity (e.g., Blatchley et al., 1997).

UV disinfection was first used at the beginning of this century to treat water in Henderson, Kentucky, but was abandoned in favor of chlorination (Bitton, 1994). UV disinfection is capable of meeting treatment requirements for wastewater reclamation and reuse (Weber, 1995; Heath, 1995; Anon., 1995). Owing to technological improvements, UV disinfection is increasingly popular, particularly in Europe and North America.

UV radiation causes damage to microbial nucleic acids and proteins at wavelengths in the range (*ca.*) of 200 nm to 300 nm. The rates of photochemically induced changes are a function of wavelength, and are generally greatest at approximately 260 nm. A particularly important form of UV-induced damage is thymine dimerization, which inhibits deoxyribonucleic acid (DNA) replication and effectively inactivates microorganisms. UV can provide efficient inactivation of bacteria, viruses, and protozoan (oo)cysts in potable water without production of substantial quantities of DBPs. Also, there is no need for handling and storage of dangerous chemical disinfectants and less space is required (Bitton, 1994).

In UV-based processes, disinfection efficacy is governed by UV dose, which is commonly defined as the product of radiation intensity and exposure time. A more formal definition of UV dose is the time-integral of radiation intensity to which an organism has been exposed. This expression of UV dose provides a realistic representation of microbial (particle) UV exposure history (Blatchley and Hunt, 1994). Regardless of definition, it is difficult to determine UV dose in real applications. Many variables can affect dose delivery in a UV disinfection system, including suspended particles, chemical oxygen demand (COD), UV-absorbing materials, reactor geometry, and hydrodynamics. With regard to optical behavior, many organic compounds may absorb radiation at germicidal wavelengths, and thus cause interference with UV transmittance in water.

In all disinfection processes, the potential exists for sub-lethally damaged organisms to repair damage following disinfectant exposure. In addition, regrowth of the microbial community will depend on source water quality, receiving water quality, and the nature and extent of disinfectant application. Therefore, the risks of microbial exposure associated with the application or omission of disinfection in a wastewater treatment system must incorporate the potential for microbial repair and regrowth.

1.2 Statement of Problems

It is evident that disinfection can effectively remove or inactivate many known pathogens; however, other concerns have been identified regarding wastewater disinfection. A major concern associated with wastewater disinfection, particularly chlorination/dechlorination, is the formation of DBPs, which may express acute or chronic toxicity toward organisms in receiving water. Other limitations of these processes are their ability to inactivate protozoan and viral pathogens and other emerging pathogenic microorganisms, as well as repair and recovery of sub-lethally injured microorganisms following disinfection. In many developed countries outside of North America, wastewater disinfection is practiced only in situations where a direct threat to human health is evident, such as discharges to bathing areas or shellfish breeding areas. The occurrence of waterborne diseases in these areas is not substantially different from North America, where wastewater disinfection has become a routine, though often seasonal, operation. These concerns have prompted reevaluation of the necessity and the risk of wastewater disinfection based on long-term effects of microbial quality and exposure of possible DBPs.

To characterize the effects of wastewater disinfection on human health from the standpoint of human exposure to bacteria, two factors should be considered. First, disinfection efficacy of a process relative to pathogenic bacteria, and their stability after disinfection, must be characterized. Second, it is necessary to characterize changes in microbial ecology after disinfection.

Microbial quality is generally monitored using indicator microorganisms (usually bacteria), since potential pathogens are quite numerous, and often difficult to quantify, due to problems associated with culturability and/or low microbial abundance. However, disinfection will not only inactivate pathogenic microorganisms but also non-pathogenic microorganisms, and different disinfection processes will have different selectivity to microorganisms based on their respective inactivation mechanisms. Under some circumstances, it is possible that pathogenic microorganisms may be more resistant to some disinfectants than indicator microorganisms. Also, recovery/repair of pathogenic microorganisms may be independent of indicator microorganisms, and may occur at vastly different rates. Therefore, the use of indicator microorganisms to characterize effluent microbial quality can lead to inaccurate representations of risk.

Coliform bacteria and enterococci are probably the most commonly used indicator organisms for characterization of microbial quality in water samples. While these bacteria have some desirable attributes in terms of their application as indicator organisms, they also have some important drawbacks. Most notable is their relative sensitivity to most commonly applied disinfectants. Coliforms and enterococci are generally more easily inactivated than most other microbial pathogens. This generalization is particularly valid when evaluating the relative sensitivity of these bacteria compared with the viral and protozoan organisms that are believed to represent the greatest risks to human health.

Another risk related to wastewater disinfection is formation of DBPs. Many DBPs have been identified in chlorination processes and found to be harmful to humans. Among the various DBPs, many are formed by reactions between oxidizing disinfectants (e.g., chlorine) and organic compounds. Largely as a result of limitations of contemporary analytical techniques, only limited information about reactions between chlorine and organic-N compounds is available. This lack of information regarding the chemistry of interactions between chlorine and reduced organic nitrogen may be partly attributable to the fact that organic chloramines have been shown to have very little ability to disinfect. Recent studies have indicated that the reactions between chlorine and organic-N compounds might be faster than chlorine and inorganic-N compounds (Wolfe et al., 1985). In chlorination of wastewater without nitrification, combined chlorine would be the predominant chlorine form; under most circumstances, combined chlorine will exist as a mixture of inorganic and organic chloramines and related compounds. The presence of organic chloramines might also lead to overestimation of disinfection efficiency since most known organic chloramines have little or no disinfection ability (Feng, 1966; Esvelt et al., 1971; Sung, 1974; Donnermair and Blatchley, 2003).

Although organic chloramines are generally ineffective as disinfectants, they may still exert acute or chronic toxicity. Studies indicate that some organic chloramines may be recalcitrant to dechlorination by sodium thiosulfate; wastewater effluent toxicity may remain after dechlorination (Helz and Nweke, 1995).

Ideally, risk assessments of wastewater disinfection should consider microbial and chemical quality. For microbial quality, many organisms are able to recover/regrow after disinfection and thus cause instability of disinfected water quality. However, a disinfection process might also cause the formation of undesirable DBPs. These factors should be taken into account, and it is critical to find balance among these factors to determine the effectiveness of a disinfection process. However, quantitative risk assessments based on chemical exposure are difficult to conduct for an aquatic matrix that is as complex and potentially variable as wastewater effluent, particularly when it is allowed to mix with a receiving water.

1.3 Objectives

This research was designed to examine the chemical and microbiological risks associated with the application of conventional wastewater disinfectants. These risks were characterized in the context of two important questions:

- Should municipal wastewater effluents be disinfected prior to discharge?
- When disinfection is necessary, how should it be accomplished?

To address these questions, two commonly applied disinfection processes (chlorination/dechlorination and UV irradiation) were applied in bench-scale experiments to effluent samples collected from several municipal wastewater treatment facilities. Facilities were selected to provide a spectrum of effluent quality, particularly with regard to N and nitrogenous compound removal in treatment. Disinfectants were applied at doses that were sufficient to accomplish compliance with relevant effluent discharge regulations based on microbial quality and chemistry. Bacterial and viral responses to disinfectants were then characterized.

Bacterial quality was monitored for six days after disinfection to investigate long-term effects of wastewater disinfection based their regrowth potential, (aerobic) respiration activity, and microbial community change. These efforts were aimed at providing detailed information regarding the stability and quality of treated samples from these two disinfection processes.

Additional bacterial assays were conducted to assess the nature and extent of damage among bacteria subjected to disinfectants. Specific objectives of these assays were to determine the relative and absolute abundance of fecal coliforms (FC), enterococci and total culturable bacteria (TCB) in treated wastewater samples, as well as the relative and absolute effects of chlorination/ dechlorination and UV irradiation. Several non-culture-based methods were applied for assessment of disinfection efficacy and to compare the results of these measurements with traditional culture-based methods.

Among waterborne microorganisms, viruses are generally believed to represent a more substantial threat to human health than bacteria. Unfortunately, viruses that express pathogenicity toward humans are generally more difficult to assay than common bacterial indicators. However, common phages share many important characteristics with human viruses, and the methods used to quantify infective phage populations are relatively easy to apply. Therefore, assays of phage responses to chlorination and UV irradiation were conducted.

The literature indicates that UV irradiation, applied at doses and wavelengths relevant in disinfection applications, results in minimal chemical change to most chemical constituents; however, some minor changes in inorganic- and organic-N may occur. The changes that occur in chlorination/dechlorination systems may be expected to involve a much broader range of compounds. However, there is evidence that chlorination of organic-N compounds may yield environmentally relevant compounds. Also, among the facilities selected for sampling, it is reasonable to expect a broad range in organic-N composition. Because the spectrum of organic-N compounds found in municipal wastewater effluents is so large, and because these compounds are generally present at low concentrations, the focus of experiments involving chemical characterization was placed on chlorination of model organic-N compounds.

The chemistry of reactions between free chlorine and several model organic-N compounds was investigated using membrane introduction mass spectrometry (MIMS). The MIMS system allowed for detailed characterization of reaction mechanisms, as well as detection and identification or reaction products. Chlorinated solutions were also subjected to toxicity assessments using a Microtox system and a respirometry assay.

Based on the results of experiments conducted to quantify microbial responses to disinfectant exposure for the conditions used in this research, a quantitative assessment of risks associated with viral exposure was conducted. Using a scenario designed to represent body contact with treated water in a surface water supply that receives municipal wastewater effluent, the risks of disease transmission associated with circumstances of no disinfection, chlorine-based disinfection, and UV irradiation were estimated and compared.

The collective results of these experiments and simulations were summarized. This information was used to address the basic questions that formed the basis of this research;

namely: should wastewater disinfection be practiced? And under the assumption that the answer to this questions is (at least sometimes) yes, how should it be practiced and why?

CHAPTER 2.0

LITERATURE REVIEW

2.1 Chlorination

2.1.1 General Information

Chlorine was first used as a disinfectant for municipal wastewater treatment in the Jersey City, NJ, Boonton reservoir in 1908 (Cheremisinoff and Cheremisinoff, 1993). Since then, chlorination has become the most prevalent wastewater disinfection process in the United States. However, chlorination is also known to result in the formation of various harmful DBPs, including THMs, HAAs, and a spectrum of other compounds (Rebhun et al., 1997). In addition, low concentrations of residual chlorine (free and combined) express toxicity toward fish and other aquatic organisms. For this reason, the U.S. Environmental Protection Agency (U.S. EPA) has established a criterion for total residual chlorine of <0.16 μM (0.011 mg/L as Cl_2) in wastewater effluent (U.S. EPA, 1984). The most common process for control of effluent residual chlorine is dechlorination through the application of reduced sulfur compounds. Since sulfur-based dechlorination processes are generally instantaneous, the wastewater is typically discharged immediately post-dechlorination.

Although used to accomplish different treatment objectives, the disinfectants that are most commonly applied in treatment of municipal wastewater (chlorine and UV radiation) have also been used extensively in the production of potable water. Although important differences between the applications of these processes exist, significant similarities also exist. Therefore, this literature review includes a number of references to material that could be considered to have a drinking water focus.

2.1.2 Concerns

Chlorination/dechlorination can cause adverse health effects in humans due to the formation of DBPs. However, there are other concerns about chlorination/dechlorination of organic-N compounds, including:

- Organic-N represents a larger percentage of total-N than inorganic-N in untreated domestic wastewater and treated wastewater effluents (Tchobanoglous and Burton, 1991). For treated domestic wastewater effluent, Isaac and Morris (1983) estimated that free amino-N ranges from 200 μg/L to 750 μg/L as N. Thus, it is possible for the concentration of organic chloramines in chlorinated wastewater effluent to be higher than the concentration of inorganic chloramines.
- Although dechlorination may be considered instantaneous and complete for inorganic chloramines and free chlorine, studies have shown that many organic chloramines or chloroorganic-N compounds are resistant to traditional dechlorination processes based on reduced sulfur. For example, chlorination/dechlorination experiments on freshly collected, tertiary-treated municipal wastewater have shown a fraction of the organic-N chloramines were dechlorinated slowly by sulfite, with half-lives of longer than 20 minutes (MacCrehan et al., 1998).

- Residual toxicity toward aquatic organisms has been demonstrated in chlorinated/ dechlorinated wastewater effluent. Toxicity responses reported in these studies, possibly caused by organic chloramines, included mortality and reduced productivity (Lenwood et al., 1981; Rein et al., 1992; Burton and Fisher, 1995; Blatchley et al., 1997).

- Although some studies have indicated limited ability of some organic chloramines to inactivate bacteria (William et al., 1985; Gowda et al., 1986; Elder et al., 1987; Swango et al., 1987), organic chloramines have been shown to be generally ineffective disinfectants (Donnermair and Blatchley, 2003).

- Most organic chloramines or chloro-organic-N compounds formed during chlorination have not yet been identified, and relatively little is known about their characteristics and fate in environment.

- Chlorination results in the production of N-nitrisodimethylamine (NDMA), which has been classified as a probable human carcinogen by the U.S. EPA. For example, studies have indicated that NDMA can be formed by reaction of NH_2Cl with dimethylamine (Mitch and Sedlak, 2002; Choi and Valentine, 2002).

Based on these and other research findings, the following generalizations can be made:

1) Chlorination of organic-N compounds can yield organic chloramines and other by-products. The reactivity and stability of the products of these reactions depends on reactant chemical structures (Hureiki et al., 1994) and available chlorine concentrations.
2) Disinfection efficacy of organic chloramines depends on chemical structure, molecular size, and polarity. It is also possible that chloroorganic-N compounds may produce toxic intermediates or mutagens (Lu and Lederberg, 1976) during microbial metabolism and exert long-term effects on the microbial ecology of receiving waters.
3) Fast oxidation reactions will consume residual inorganic chlorine quickly, and this may be important in diminishing disinfection efficacy in wastewater containing organic-N compounds.

2.1.3 Chlorination of Organic-N Compounds

When organic-N compounds are subjected to chlorination, they may undergo two types of reactions: substitution and oxidation (Figure 2-1). Substitution reactions will generally result in the formation of organic chloramines and other chlorinated by-products, while oxidation will lead to the formation of oxidized or fragmented compounds, including carbon dioxide and chloride.

Most research regarding chlorination of organic compounds has focused on the identification and kinetics of volatile and semi-volatile by-products; consequently, only limited information is available regarding organic chloramines and their formation. This may be because they are generally believed to have much less or even no ability to inactivate microorganisms, complexity, and lack of analytical methods. However, recent research has identified possible concerns regarding the effects of organic-N compounds on human health and the environment.

Past research has illustrated the limited ability of (most) organic chloramines to inactivate waterborne microorganisms; however, some studies have indicated that some organic chloramines may demonstrate quantifiable antimicrobial efficacy. Compounding this problem is the inability of common analytical techniques (usually based on titration or spectrophotometry) to differentiate inorganic and organic chloramines. To address this problem, MIMS has been applied for differentiation and quantification of inorganic chloramines in aqueous solution (Shang et al., 2000). This technique has also been applied successfully for online monitoring of volatile DBP formation during the chlorination process (Chang and Her, 2000; Rios et al., 2000; Shang et al., 2000).

2.1.3.1 Source and Concentration

A great number of amines are used in industrial chemicals, agricultural applications, and found as constituents of the natural environment. Common applications of aliphatic amines in industry include corrosion inhibitors, desulfuration reagents in petroleum refineries, antioxidants, vulcanization accelerators, plasticizers, pharmaceuticals, detergents, dyes, textiles, pesticides, surface-active agents, and dehydrohalogenation catalysts (Astle, 1961). Besides industrial sources, amines can come from decomposition of organic matter, especially protein wastes. Rendering plants and packing houses are important sources of these substances.

Vallentyne (1964) stated that α-amino N has been found to comprise as much as 60% of the total organic-N in freshwaters. More recent data (Issac and Morris, 1983) revealed median values for the ratio of organic-N/ total Kjeldahl nitrogen (TKN) in secondary effluent and nitrified effluent to be 0.59 and 0.92, respectively.

Ram and Morris (1982) summarized the occurrence of N compounds and listed several amino acids excreted by humans. Based on these limited data, representative concentrations of amino N, associated with only amino acids in domestic wastewater, were estimated to be between 0.3 mg/L and 1 mg/L as N. For comparison, the average ammonia-N concentration in non-nitrified secondary effluents reported by Ram and Morris (1982) was 11.6 mg/L (as N). Thus, while the concentration of organic N can be a relatively large fraction of the reduced N in effluent, amino-N may comprise ≥10% of the ammonia-N concentration in a nonnitrified secondary effluent. The concentration of organic amines in industrial wastewater varies depending on type of industry. The concentrations of N-containing odorous compounds detected in wastewater treatment plant (WWTP) samples are listed in Table 2-1 (Hwang et al., 1995).

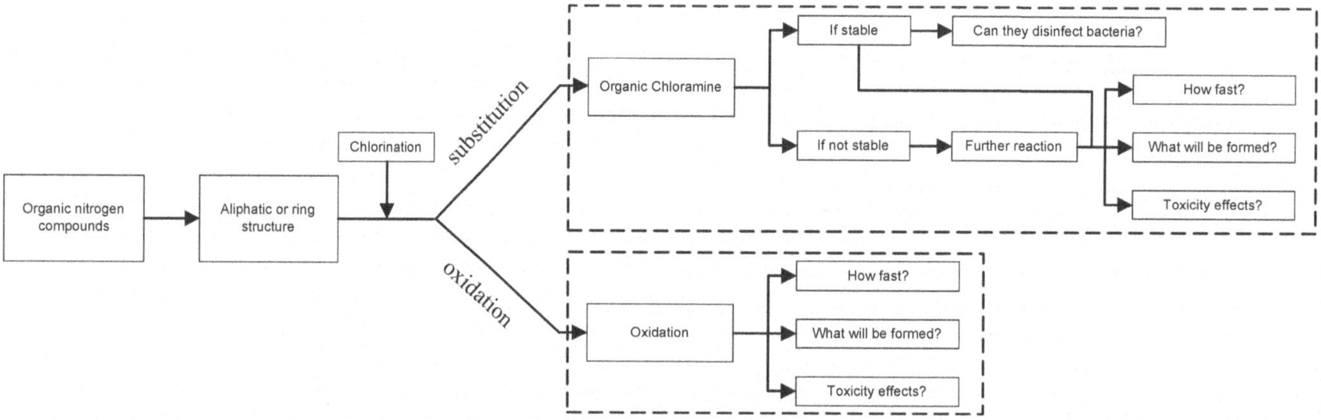

Figure 2-1. Roles of Organic-N Compounds in the Chlorination Process. Chlorination of organic-N compounds involves two major reaction types: substitution and oxidation; both can lead to the formation of DBPs.

Table 2-1. Nitrogen-containing Odorous Compounds Detected in WWTP Samples (unit: μg/L) (Hwang et al., 1995).

	Trimethylamine	Dimethylamine	n-propylamine	Indole	3-Methylindole
Influent	78	210	33	570	700
Primary effluent	52	215	31	430	640
Secondary effluent	36	230	28	—[b]	—
Anthracite[a]	17	180	13	—	—
Activated carbon	16	165	9	—	—

[a]Effluent of advanced treatment aerated biofilter
[b]Below the detection limit

2.1.3.2 Concerns

Amines can cause hygienic problems in the form of malodor, as well as irritation of the skin, mucous membranes, and eyes. An important health hazard of amines relates to the formation of potentially carcinogenic nitrosamines in the presence of nitrite. When amines are present in wastewater, biodegradation is generally considered to be the most economical method of treatment. However, due to enzymatic deficiencies or their toxicological properties, some amines are biorecalcitrant and may show inhibition in biological wastewater treatment systems (Eilersen et al., 1994; Moore and Baker, 1996). Further, when amine-laden wastewater is imposed on biological treatment plants, the performance of these facilities may be challenged. Amines can result in low sludge settleability or sludge bulking (Chong, 1994). Amines can be important components of organic-N in wastewater effluents.

Amines can cause problems not only in biological treatment, but also in chlorination/dechlorination. The kinetics and mechanisms of chlorine transfer from chloramines to amines in aqueous media have been investigated (Issac and Morris, 1983; Ferriol et al., 1991). Some chlorinated amines are resistant to dechlorination with S(IV). Helz and Nweke (1995) indicated that dechlorination with S(IV) can remove 87% to 98% of residual chlorine but a remainder, which may exceed regulatory limits, is very slowly reduced. Judging from kinetic evidence, chlorinated secondary organic amines and peptides probably contribute to S(IV)-resistant residual chlorine. Furthermore, most of the S(IV)-resistant fraction was extractable into octanol. Its hydrophobic character suggested that this fraction could be harmful to organisms in receiving waters (Krasner, 1989).

2.1.3.3 Analytical Methods

Several organic chloramines, such as chloroglycine, chlorovaline, chlorodimethylamine, and N-chloromethylamine have been identified as chlorination by-products in wastewater, based on the results of LC and GC retention time, as well as the molecular ion and fragmentation patterns in GC/MS analysis followed by dansyl ($ArSO_2^-$) derivatization (Choshen et al., 1987).

The analysis of amines, especially by GC/NPD, TEA and MS, is well known and often described in the field of air, clinical and food analysis (Vatsala et al., 1994; Gronberg and Jonsson, 1992). However, detection limits (mg/L - μg/L, typically) are generally too high for organic trace-level analysis of surface and drinking water. Sample enrichment is necessary to reach detection limits in amine analyses, and analysis is usually achieved by derivatization methods. Direct determination of chlorination products of organic amines has been demonstrated using MIMS (Kotiaho et al., 1991 a, b).

2.1.3.4 Chemistry

2.1.3.4.1 Chlorine Transfer

The reactions between most organic-N compounds and chlorine are very fast; many amino acids have reaction rate constants for reactions with chlorine that are greater than those that lead to formation of NH_2Cl (Wolfe et al., 1985). The significance of this phenomenon is that it will affect the distribution of chlorine among the various forms of dissolved N in wastewater chlorination.

Scully and Hartman (1996) reported that wastewater samples with high organic-N concentration were generally disinfected with preformed NH_2Cl at much faster rates than similar samples disinfected with $Cl_{2(aq)}$. This may also imply that organic-N compounds have higher reaction rate constants with free chlorine than ammonia. Isaac and Morris (1983) studied the transfer of active chlorine from chloramines to nitrogenous organic compounds, including amines, amino acid, glycine derivatives, and amides. They found active chlorine transfer would progress by two approaches: 1) hydrolysis of NH_2Cl to form HOCl, which will react with remaining unchlorinated nitrogenous organic compounds; and 2) direct transfer of active chlorine from NH_2Cl to organic-N compounds. Ferriol et al. (1991) investigated chlorine transfer from chloramine to amines in aqueous solution. They found that chlorine transfer from NH_2Cl to amines can be regarded as a possible two-route mechanism: 1) the general acid-catalyzed route and involves NH_2Cl and the neutral form of amines; and 2) an interaction between NH_3Cl^+ and the neutral form of amines and this route is not kinetically separable from the first route.

2.1.3.4.2 Kinetics and Pathways

Amines can react rapidly with residual chlorine. Primary and secondary aliphatic amines react rapidly with chlorine to form organic chloramines, mainly by substitution reactions as described by reactions (2.1) and (2.2) (Margerum et al., 1978):

$$HOCl + RNH_2 \rightarrow RNHCl + H_2O \tag{2.1}$$
$$HOCl + RR'NH \rightarrow RR'NCl + H_2O \tag{2.2}$$

Primary organic monochloramines can react further with additional free chlorine to form organic dichloramines by the following reaction:

$$HOCl + RNHCl \rightarrow RNCl_2 + H_2O \tag{2.3}$$

Weil and Morris (1949) proposed two different mechanisms for this reaction in the pH range 4 to 12. One is the same as that proposed by Margerum et al. (1978). The other is a mechanism in which hypochlorite anion reacts with protonated amine. Judging from the pK_a values of HOCl and common aliphatic amines, under the normal near-neutral or slightly alkaline pH conditions of wastewater chlorination, most amines would be present in their protonated form, while HOCl would be present in neutral or anion form. Although chlorine transfer in primary and secondary amines has been well documented, there is little information regarding chlorine transfer in chlorination of tertiary amines.

After formation, organic chloramines may undergo further reactions in the presence of available chlorine. However, there is only limited information regarding these reactions. Based on the work of Pedersen et al. (1999), possible pathways for the formation of cyanide from the reaction of NH_2Cl with formaldehyde have been proposed:

$$CH_2O + NH_2Cl \rightarrow CH_2(OH)NHCl \quad (2.4)$$
$$CH_2(OH)NHCl \rightarrow CH_2NCl + H_2O \quad (2.5)$$
$$CH_2NCl \rightarrow HCN + HCl \quad (2.6)$$

Nweke and Scully (1989) and Scully and Conyers (1997) found that aldehydes can be formed from chlorination of amino acids, based on the following hypothesized pathway:

$$R\text{-}CHNH_2\text{-}COO^- + OCl^- \rightarrow R\text{-}CHNHCl\text{-}COO^- \quad (2.7)$$
$$R\text{-}CHNHCl\text{-}COO^- \rightarrow R'\text{-}CH_2CHNH \text{ (by loss of } CO_2 \text{ and } Cl^-) \quad (2.8)$$
$$R'\text{-}CH_2CHNH + H_2O \rightarrow R'\text{-}CH_2CHO \text{ (by loss of } NH_3) \quad (2.9)$$

Imines would also be formed as an intermediate (reaction (2.8)). This result corroborates the finding of Trehy et al. (1986).

There is relatively little information regarding the formation of DBPs from chlorination of amines. Among numerous DBPs that result from chlorination, cyanogen chloride, dichloroacetonitrile, trichloroacetonitrile and chloropicrin have been identified as DBPs containing N. Of these DBPs, cyanogen chloride is of particular interest because of its toxicity and because it is commonly observed (in low concentrations) in chlorinated drinking water (Pedersen et al., 1999). The World Health Organization (WHO) proposed a guideline value of 70 μg/L for all cyanide species (including cyanogen compounds) in drinking water (WHO, 1993).

2.1.3.5 Toxicity

Many substances in wastewater may display toxicological effects in humans and aquatic organisms after chlorination. These toxicity effects may come from mutagens, carcinogens, and teratogens. For example, dichloroacetonitrile (8.6 ppm) can inducc highly significant increments in the frequencies of sex chromosome loss and gain (Osgood and Sterling, 1991). Sodium cyanide has also been found to be a highly effective inducer of germline aneuploidy (Osgood and Sterling, 1991). The genotoxicity of haloacetonitrile compounds may be attributable to their high affinity towards single-stranded DNA (Nouraldeen and Ahmed, 1996).

Many methods exist for assessment of toxicity based on test purposes, species, and/or water sample (fresh water or marine sample) used. Common methods for assessing toxicity in aquatic environments include: 1) the *Daphnia* 48-hour acute toxicity test, 2) the algal 96-hour growth toxicity test, and 3) acute toxicity tests with aquatic vertebrates and macroinvertebrates.

In recent years, there has been increasing interest in using bacterial screening for toxicity assessments (Hao et al., 1996; Chen et al., 1997; Campbell and Stribig, 1999; Choi et al., 2002). Studies of effects on microbial function or activity constitute a relatively rapid and sensitive approach to measure stress caused by toxicants and may be preferred to studies of the effects of population growth. These assays can be further categorized by the response parameters used for quantification of toxicity, including: 1) assays based on monitoring transformation of carbon, nitrogen, or sulfur (Williamson and Nelson, 1983); 2) assays based on bioluminescence output (Beckman Instruments, 1982; Bulich, 1982); 3) assays based on the measurement of oxygen consumption using dissolved oxygen (DO) probes (International Standards Organization [ISO], 1984); 4) assays based on the determination of glucose uptake activity (Larson and Schaeffer, 1982) and 5) assays based on the measurement of microbial dehydrogenase enzyme activity (Liu, 1981; Liu and Dutka, 1984).

The Microtox acute toxicity test is a 15-minute metabolic inhibition test that uses freeze-dried luminescent bacteria (*Vibrio fischeri* NRRL B-11177) to assess acute toxicity of water samples. The Microtox chronic test is a highly sensitive metabolic inhibition test that also uses a strain of luminescent bacteria. Hydrated cells from the chronic test reagent, when grown in a defined medium, undergo induction of metabolic pathways (including luciferase enzymes) needed for normal metabolism, growth, and cell division. These enzymes, as well as cellular structures, become available for inhibition or disruption by toxic chemicals. The integral association of the luciferase pathway (light production) with cellular growth and metabolism provides a sensitive endpoint for measuring inhibition of cellular activity (chronic toxicity).

In respiration assays, toxicity of a chemical is characterized by measuring the difference of metabolic activity of microorganism(s) before and after exposure to toxicant. Oxidation is the main mechanism for removal of biodegradable matter. Therefore, the oxygen uptake (OUR) per unit volume and per unit time (i.e., the respiration rate), is the key variable that characterizes the metabolic activity of the microorganism. The OUR is measured using a DO probe, which measures the DO concentration. The respiration rate is then calculated from time-course changes in the DO concentration. Toxicity assessment can be expressed by inhibition of maximum respiration rate under different toxicant concentration and contact time. The maximum respiration rate is measured in the presence of excess substrate.

Many studies have shown that disinfectants can react with natural organic substances present in water supplies to form by-products with mutagenic and/or carcinogenic activity (Dolara et al., 1981; Gauthier et al., 1989; Daniel et al., 1994). DeVeer et al. (1994) examined the effects of UV exposure on toxicity responses in aqueous solutions containing pesticides and naturally occurring organic compounds. Ames tests and sister chromatid exchange tests were performed before and after UV irradiation. Test results showed no toxicity by either test before or after irradiation. Oppenheimer et al. (1997) indicated that no chronic toxicity was observed in a full-scale UV irradiated effluent; however, a decline in the reproductive rate of *Ceriodaphnia dubia* was observed for the full-scale chlorinated effluent. Blatchley et al. (1997) tested the effects of several disinfection processes (including chlorination/dechlorination and UV irradiation) on effluent toxicity using a *C. dubia* assay. Their results demonstrated that changes in toxicological response, as measured by death or lack of reproduction attributable to disinfectant exposure, were generally more severe following chlorination/dechlorination than UV irradiation. Monarca et al. (2000) tested the influence of different disinfectants, including chlorine, chlorine dioxide, ozone, peracetic acid, and UV irradiation, on the formation of mutagenic and toxic compounds in wastewater. Their results indicated that all disinfectant treatments produced bacterial mutagenicity, particularly after treatment by chlorine dioxide or ozone.

Ono et al. (1992) tested the genotoxicity of organic substances in municipal wastewater by using the *umu* test system which can detect *umuDC* genes that are known to cause error-prone repair. Their results showed genotoxic substances in secondary effluent were decomposed by ozone into non-genotoxic substances. Ono et al. (1996) also tested the influence of treatment processes on the removal of genotoxicity and found genotoxicity could not be removed by biological treatment processes and ultrafiltration; however, it was reduced by approximately 90% by coagulation-sedimentation. They also investigated the presence of carcinogenic heterocyclic amines (such as Trp-P-2 on the order of 10^{-6} mg/L) in river water downstream of several wastewater treatment facilities, and concluded that these carcinogens were derived from the effluent of wastewater treatment facilities containing human waste (Ono et al., 2000). Another study was done by Silva et al. (1999) using river water to study the potential of water contaminants to induce tissue lesions in rats. Their findings suggest that the superoxide anion elicits lipoperoxide formation and induces tissue damage.

The chlorination byproducts of organic-N compounds are of concern since dichloroacetonitrile has been found to be mutagenic by the Ames test (Bull, 1983) and tumor-initiating by the Sencar mouse skin bioassay (Bull and Robinson, 1985). In addition, consumption of chlorinated water resulted in the *in vivo* formation of dichloroacetonitrile and trichloroacetonitrile (Mink et al., 1983). The potential for *in vivo* formation of chlorination byproducts should be considered, since chlorine has been shown to react with polypeptides as well as with amino acids (Goldschmidt et al., 1927).

2.1.4 Dechlorination

Traditional dechlorination processes using sodium bisulfite are often viewed as being instantaneous and complete; the reactions between bisulfite and free chlorine, or bisulfite and inorganic combined chlorine are extremely rapid. However, less is known about the kinetics of reactions between bisulfite and organic combined chlorine. Likewise, limited information is available regarding the fate of sulfur (IV) in natural waters (Kijak and Helz, 1988). In general, effluents dechlorinated with bisulfite are discharged immediately after dechlorination. Some organic-N compounds may be recalcitrant to biodegradation (e.g., propionanilide) (Oyamada et al., 1980). Chlorination of effluent containing these compounds, which may remain in solution following biochemical treatment, leads to the formation of organic chloramines.

Some chlorinated organic-N compounds have been observed to be resistant to traditional dechlorination using S(IV) (Helz and Nweke, 1995; Jensen and Helz, 1998 a,b; MacCrehan et al., 1998). Helz and Nweke (1995) tested samples from two wastewater treatment facilities and found that dechlorination was capable of removing 87% to 98% of residual chlorine, but the remainder, which may exceed regulatory limits, was very slowly reduced. The dechlorination rate and extent are likely to depend on the structure of the organic-N precursors. Chlorinated secondary organic amines and peptides have been shown to be important contributors to S(IV)-resistant residual chlorine.

Jensen and Helz (1998a) tested dechlorination kinetics of N-chloropiperidine, a genotoxic compound that has been found in chlorinated municipal wastewater and found that the rate at which it was reduced, and thus detoxified by sulfite, was enhanced in slightly acidic pH conditions. They also showed that some organic-N-chloramines were dechlorinated slowly by sulfite with half-lives of >20 minutes. Those sulfite-resistant compounds, based on chromatographic retention and kinetic behavior, were hypothesized to have originated from peptides and proteins (MacCrehan et al., 1998). Further studies by Jensen et al. showed that the dechlorination rate constants of N-chloropeptides were 1-2 orders of magnitude smaller than those for NH_2Cl and some aliphatic organic chloramines. These kinetic data suggest that reducing N-chloropeptides in wastewater by $\geq$75% will require reaction times >30 minutes at environmentally acceptable S(IV) doses and pH values (Jensen and Helz, 1998b).

Although the toxicity effects of chlorinated organic-N compounds are not well-known, preliminary results from long-term respiration tests using wastewater with high (>10 mg/L) and low (<0.1 mg/L) ammonia nitrogen (NH_3-N) burdens have demonstrated a distinct difference in oxygen consumption. In wastewater, high NH_3-N conditions will generally correspond with high organic-N; this suggests that organic-N compounds, after chlorination, might exert a long-term effect on respiration behavior.

Because of the potential problems caused by chlorination of organic compounds, removal of these compounds may represent an important issue in some wastewater systems. Previous research has focused on the elimination of chlorinated DBP precursors. The most popular processes used for removal of organohalogen and organohalogen precursors are granular activated carbon (GAC) adsorption and advanced oxidation (Bauman and Stenstrom, 1990a,b). These processes can be effective in reducing THM production during chlorination; however, relatively little information has been obtained regarding the effectiveness of these processes in removal of non-volatile organohalogen compounds. In addition, these processes also have important technical and economic limitations. GAC columns may be subject to degradation, channeling, and premature chlorine breakthrough. The GAC may also become a bacterial breeding ground, especially within the chlorine-free, but nutrient rich, lower areas of bed, with subsequent increased bacterial counts noted downstream. Advanced oxidation may require that additional chemicals be used, thereby requiring onsite chemical storage, handling procedures, and precautions. Also, the approach adds an additional substance, with possible by-products, to an already complex water purification process. Alternative processes for elimination of organohalogen compounds include pre-sonication (Ma and Lin, 1998) and UV irradiation (Belhacova et al., 1999; Saito et al., 1992).

Lind (1995) listed and compared several alternative options for wastewater dechlorination. These alternatives included: sulfur dioxide, sodium metabisulfite, sodium bisulfite, sodium sulfite, sodium thiosulfate, ammonium bisulfite, and ammonium thiosulfate. He also concluded that the choice of dechlorination agent should depend on site-specific needs, for example: safety, convenience, cost, and operation requirements.

Studies of aqueous free chlorine removal by sunlight (tropospheric solar radiation) have shown that the photolysis half-life of aqueous free chlorine at pH = 10 was 12 minutes when measured at the surface. When irradiating the sample with UV radiation (254 nm, 3.7 μEinstein/L·s), the half-life was estimated to be 10.3 minutes (pH 10) and 10.9 minutes (pH 5.4) (Nowell et al., 1992a). During the photolysis of $Cl_{2(aq)}$ by UV radiation, reactive free radicals (HO• and Cl•) will be formed. HO• production varied strongly with pH and slightly with the wavelength of irradiation. Nowell and Hoigne (1992b) also reported that OCl^- yielded 0.1 HO• at pH of 8, and HOCl yielded 0.9 HO• at pH 5. Combs and Ammerer (1998) reported the extinction dosage for removing 0.1 mg/L of free chlorine to be 0.27 mJ/cm^2 by using a single, low-pressure (LP) monochromic UV lamp with wavelength of 254 nm, and the extinction dosage for removing 1 mg/L of free chlorine was estimated to be 1.83 mJ/cm^2.

2.2 UV Disinfection

2.2.1 General Information

The limitations of chlorine-based disinfection have motivated the consideration of alternative disinfection processes that are more friendly to human health and the environment. The most promising of these alternatives may be UV disinfection because it is capable of reducing the concentration of viable pathogenic microorganisms and generally produces few DBPs.

Several water characteristics are important to UV disinfection, including UV transmittance and particulate matter. Respectively, these attributes are responsible for absorption and scattering of incident UV radiation. Of the dissolved compounds, organic- and inorganic-N can play important roles as absorbers. As a result, disinfection efficacy of UV irradiation will be reduced by the presence of these constituents (Bitton, 1994). Furthermore, the lack of standardization in determining UV dosage in UV systems makes accurate design very difficult. For the same reason, it is often difficult to compare UV disinfection studies on the basis of "dose".

Collimated beam experiments should be a component of UV system design (Blatchley, 1997). However, real full-scale inactivation behavior cannot be deduced solely from collimated beam tests because continuous-flow UV systems are characterized by a dose distribution (Blatchley and Hunt, 1994; Chiu et al., 1999 a,b; Lyn et al., 1999; Lyssandridou et al., 2002).

UV radiation, when combined with a photocatalyst such as titaniaum oxide (TiO_2), has also been shown to be capable of inactivating microorganisms. The fundamental behavior of a photocatalytic reactor is different from that of a pure photochemical reactor. For example, Belhacova et al. (1999) tested the inactivation of *Escherichia coli* (strain DH5α) and bacteriophage (λNM1149) under a flow-through photoreactor with an immobilized TiO_2 layer and found that the inactivation kinetics were approximately first-order and the initial rate depended on the radiation intensity. The mechanism of inactivation was hypothesized to be the destruction of the surface structure of bacterial cells caused by hydroxyl radicals. By contrast, in a purely photochemical system used for disinfection, the mechanism of inactivation involves photochemical alteration of nucleic acids and proteins. The exterior surface structures of microorganisms are largely unaffected by UV. Saito et al. (1992) described the inactivation of *Streptococcus sobrinus* in an aqueous suspension of TiO_2 subjected to UV irradiation. Cell membranes appeared to be damaged by the action of reactive oxygen species, and the changes in membrane permeability led to leakage of potassium ions, protein, and ribonucleic acid (RNA) molecules, causing loss of cell viability. Cell walls were also decomposed. Matsunaga et al. (1988) observed complete dimerization of coenzyme A (CoA) under UV irradiation when incubated with TiO_2 suspensions, suggesting a relationship between the amount of dimerized CoA and viability loss of bacteria.

For the design and characterization of a UV reactor, tremendous advances have been made in recent years. Torrentera et al. (1994) presented a method to characterize a UV disinfection system in which radiochromic dye films were used to evaluate the uniformity of radiant exposure within the system. Blatchley and Hunt (1994) used the results of point source summation and full-scale bioassays applied to several full-scale UV systems to characterize the delivery of UV radiation to microorganisms. On the basis of this information and further analysis of the fundamental physical behavior of photochemical reactors, the concept of a UV dose distribution was introduced and the effects of several lamp usage patterns on bacterial inactivation were also reported. Chiu et al. (1999a,b) presented an integrated model for assessing disinfection efficacy of a flow-through UV disinfection system. In this model, the UV intensity field and simulated, individual particle trajectories were considered for estimation of the dose distribution delivered by a continuous-flow UV system. In turn, predictions of microbial inactivation were developed through integration of the dose distribution and microbial dose-response behavior (as measured using a collimated beam), using an approach that was mathematically identical to the so-called "segregated flow" model (Levenspiel, 1972).

2.2.2 Inactivation Theory

Microbial inactivation in a UV reactor is brought about by photochemical changes to nucleic acids and proteins (Setlow, 2002). This is indicated by similarities between the absorption spectra of nucleic acids and the action spectra for cell inactivation (Moan, 1989). However, in some cases microbial action spectra are slightly red shifted compared with the DNA absorption spectrum, suggesting that absorption of UV photons by proteins also plays some role. Action spectra of UV damage to physiological functions directly related to proteins, such as sodium and amino acid transport, closely resemble the absorption spectra of proteins. For UV radiation with wavelengths up to 365 nm, cyclobutyl pyrimidine dimers (CPDs) are the most frequent lesions as far as total DNA is concerned. Formation of other photoproducts is also possible in certain specific DNA sequences (e.g., pyrimidine (6-4) pyrimidone photoproducts). CPDs induced by UVC result in a reduction of the synthesis of high molecular weight DNA, while subsequent removal of CPDs by exposure to photoreactivating light will cause inhibition of the synthesis of low molecular weight DNA. Other inactivation mechanisms include single-strand breaks (SSBs), double-strand breaks (DSBs) in DNA, and DNA protein cross-links (DPCs). However, at 254 nm, SSBs and DSBs are produced two to three orders of magnitude less frequently than are CPDs and probably do not play any significant biological role.

Another advantage of UV irradiation is its ability to inactivate some viruses and protozoan (oocysts), such as *C. parvum*, that are highly resistant to conventional chlorine-based disinfectants. Mofidi et al. (2001) demonstrated the inactivation of *C. parvum* oocysts by UV radiation. For a pulsed UV dose of 16 mJ/cm^2, more than 2 $\log_{10}$ inactivation of oocysts suspension was achieved. No statistical difference in disinfection power was found between the medium-pressure (MP) and pulsed-UV lamps. Results from both lamps showed that UV doses as low as 7.5 and 11 mJ/cm^2 provided inactivation of infectious oocysts of *C. parvum* at 1 $\log_{10}$ (90%) and 2 $\log_{10}$ (99%), respectively.

2.2.3 Repair/Regrowth

An important consideration in the application of UV disinfection is the potential for regrowth/repair following disinfection. After DNA damage by UV exposure, many microorganisms are equipped with repair enzymes and repair pathways, some of which operate on UV-induced lesions. There are three major mechanisms associated with DNA repair: 1) excision repair; 2) photoreactivation; and 3) repair of DNA SSBs and DPCs.

Excision repair is probably the most important mechanism of repair of UV-induced DNA damage owing to its speed and precision. Studies have shown that > 60% of the CPDs were removed from the actively transcribed gene within four hours after the irradiation, while only 25% of the CPDs were removed from the overall genome within the same time (Bohr et al., 1986; Mellon et al., 1986). This rapid repair of active genes is in agreement with the rapid resumption of RNA synthesis after UV irradiation (Mayne and Lehmann, 1982).

Photoreactivation is usually achieved by photoreactivating enzymes (PRE); however, both CPDs and (6-4) photoproducts can be nonenzymatically photolysed. More important is enzymatic photoreactivation which specifically splits CPDs but not (6-4) photoproducts. PRE may have other functions than catalyzing the photosplitting of CPDs since it is also found in animals that live in complete darkness and in organs, such as frog liver, that are not exposed to light. Some studies have indicated that the repair of (6-4) photoproducts appears to be faster during exposure to visible light than in darkness, even though PRE does not act directly on these lesions (Eggset et al., 1983, 1987; Rosenstein, 1988).

Kashimada et al. (1996) assessed photoreactivation of FC following UV disinfection. Inactivation of FC was observed simultaneously during photoreactivation by sunlight. The photoreactivation rate increased with increasing visible light intensity at 360 nm. However, the maximum survival value may not be affected by visible light intensity. Even without sunlight reactivation, bacterial regrowth in seawater was observed after UV irradiation of sewage effluents (Lazarova et al., 1998). The maintenance of respiratory activity after disinfection could explain the increase in culturable cells in receiving systems.

Repair and regrowth can also occur following chlorination. Studies indicated the occurrence of sublethal injury to coliforms during chlorination (Jopling, 1975; Seidler et al., 1977; Olson, 1982; LeChevalier et al., 1987; Tosa et al., 1995). It is possible that sublethally injured microorganisms may subsequently recover, resulting in regrowth and deterioration of water quality. El-Rehaili (1995) observed increased biochemical oxygen demand (BOD) and COD in wastewater effluent after chlorination, which might be used as assimilable substrate for bacteria following chlorination/dechlorination. Schmidt et al. (1998) studied the effects of chlorination on bacterial regrowth potential (BRP) by using algogenic organic matter (AOM). Their results indicate that the chlorination of intact algal cells may cause a significant increase of the biodegradability in the water if residual chlorine is consumed completely. Moreover, in the case of chlorination of the AOM-containing waters with chlorine dioxide, the relative increase of the BRP was lower in comparison to the chlorinated waters. Yeh et al. (1998) indicated that in a sample with assimilable organic carbon (AOC) concentration of >110 μg acetate-C/L, heterotrophic plate count values were significantly higher than those with lower AOC, even when free chlorine residual was maintained at >1.0 mg/L in drinking water sample. Ng et al. (2001) indicated that a residual chlorine concentration of 0.2 mg/L as Cl_2 may not be sufficient to inhibit regrowth of bacteria in industrial water with a significant AOC level, and an AOC concentration of 30 μg to 77 μg acetate-C/L in reverse osmosis product water demonstrated a potential of biological instability without disinfection.

2.3 Factors Affecting Wastewater Effluent Quality

2.3.1 Impacts of Upstream Treatment Process

Different upstream treatment process will result in different characteristics of effluent quality. For example, biologically nitrified wastewater without denitrification usually has a high concentration of nitrate, as high as 10 mg/L to 30 mg/L (as N). By comparison, <1 mg/L (as NO_3-N) would be expected for wastewater subjected to nitrification/denitrification. High concentrations of organic-N in the effluent may exert a risk of effluent toxicity if chlorination/ dechlorination is chosen as a disinfection process, due to the formation of organic chloramines. Similarly, process behavior can have a profound effect on effluent quality in terms of particulate composition. Suspended solids will adversely affect disinfection process performance when UV or chlorination/dechlorination are applied.

Upstream treatment will also affect microbial composition. Chauret et al. (1999) indicated that during aerobic wastewater treatment (excluding sludge digestion), *C. parvum* and *Giardia lamblia* were reduced by 2.96 $\log_{10}$ and 1.4 $\log_{10}$, respectively. *Clostridium perfringens* spores, *C. perfringens* total counts, somatic coliphages, and heterotrophic bacteria were reduced by approximately 0.89 $\log_{10}$, 0.96 $\log_{10}$, 1.58 $\log_{10}$, and 2.02 $\log_{10}$, respectively. All other microorganisms tested were reduced by at least 3.53 $\log_{10}$. During anaerobic sludge digestion, no statistically significant reduction was observed for *C. perfringens*, *Enterococcus* sp., *Cryptosporidium* oocysts, or *Giardia* cysts. Watanabe et al. (1997) found bacterial counts for FC groups, enterococcus and *Salmonella* were substantially decreased after thermophilic digestion, as compared to a mesophilic process. Almasi and Pescod (1996) studied the mechanisms of pathogen removal in anoxic wastewater stabilization ponds and found that volumetric organic loading, temperature, light, pH, DO and oxygen reduction potential (ORP) will affect the removal efficiency of pathogens. Rose et al. (1996) studied removal of pathogenic and indicator microorganisms by a full-scale water reclamation facility and indicated that the combination of biological treatment, sand filtration, and chlorination employed at the plant reduced viable total and FC by >7 log_{10} and coliphages and enteroviruses by >5 log_{10}. Protozoa (*Cryptosporidium* and *Giardia* spp.) were reduced by >3 log_{10} and bacteriophage were decreased by 1.5 log_{10} to 1.6 log_{10}. They also found indicator and pathogenic microorganisms reduced by anaerobic treatment by approximately 1 log_{10}.

Several studies also suggested that the efficiency of chlorination is strongly affected by nitrification (organic, ammonia, nitrite, and nitrate concentrations), and in particular in presence of nitrite or lack of ammonia. Calmer et al. (1998) provided an analysis of the literature and guidance for better operation of chlorination/dechlorination with nitrification related problems. Furthermore, Rebhun et al. (1997) have shown that the potential of dissolved organic halogen formation is up to four times lower in completely nitrified effluents compared to ammonia-containing effluents.

2.3.2 Impacts of Disinfection Process

Disinfection processes will each have a characteristic impact on the microbial community because of their different inactivation mechanisms. Lazarova et al. (1999) reviewed several disinfection processes (chlorination/dechlorination, UV irradiation, ozonation, and membrane filtration) in terms of efficacy, regrowth potential, design parameters, and practical applications and concluded that the choice between these disinfection technologies depends on wastewater quality, standards, specific reuse applications, and wastewater treatment works capacity.

2.4 Impacts to Receiving Waters

Receiving water quality will affect long-term disinfection effectiveness in terms of regrowth and microbial ecology. Important factors in the receiving waters include: 1) nutrient composition; 2) nutrient concentration; 3) dilution (ratio of $F_{receiving\ water}/F_{effluent}$, where F = flow rate); 4) presence of other possible nonpoint and/or point sources of contamination; 5) meteorological factors (e.g., solar intensity) (Holm-Hansen et al., 1993; Acher et al., 1997; Yang et al., 2000) and hydrodynamic conditions; 6) ecological conditions; and 7) predators (Stott, 2002). The factors listed above are site-specific and difficult to control, yet they may be the most important factors that affect the long-term impact of a disinfected effluent.

Conversely, disinfected effluents can have important impacts on receiving water ecosystems. Although dechlorination is applied following chlorination to reduce the toxicity of chlorinated effluent, Helz and Nweke (1995) indicated that some compounds that comprise residual chlorine, exceeding common regulatory limits, may remain in dechlorinated effluent. Since sodium bisulfite is usually added as a dechlorination reagent; DO consumption might also have some impacts on the receiving waters, sometimes additional aeration is required before discharge.

Osmotic stress to receiving waters (e.g., seawater) also plays an important role in bacterial metabolism (Oliver, 1992). In some cases this stress combined with disinfection does not permit fecal bacterial regrowth (Chan and Killick, 1995).

2.5 Bacterial Pathogens

The bacteria of public health concern are the true and opportunistic pathogens that have a high likelihood of producing disease pathologies in humans. The most commonly referred to pathogens include *E. coli*, *Salmonella* sp. and *Shigella* sp. Unfortunately, these bacteria are difficult to isolate and identify, and are too costly to monitor for on a routine basis. Therefore, federal and state regulations require that wastewater treatment facilities monitor for groups of bacteria whose presence and abundance is supposedly correlated with the presence of bacterial pathogens.

FCs have been used since the mid-1900s as indicators of microbiological quality in natural and treated waters. However, recent data suggests that when monitoring marine waters FCs may not be appropriate as their survival rates are significantly reduced when compared to fresh water rates (Davis and Valentino, 1985; Rees et al., 1998; Rozen and Belkin, 2001; Sinton et al., 1994; Solic and Krstulovic, 1992).

The research on which wastewater effluent discharge regulations are based was performed using culture-based methods. However, recent research has shown that laboratory-grown bacterial cells are substantially more susceptible to disinfectants than indigenous cells (LeChevallier et al., 1988; Lisle et al., 1998; Matin and Harakeh, 1990; Olson and Stewart, 1987; Stewart and Olson, 1992).

Recent progress in the development of in situ, intracellular fluorescent probes has facilitated the rapid assessment of bacterial physiology in biofilm and planktonic populations without cultivation. Efforts to describe the physiological activities of bacterial populations have been hindered by the inability of culture-based techniques to recover a significant fraction of bacteria present in a sample (Amann et al., 1995) and by traditional dependence on methods that yield community averages of physiological indices, such as oxygen consumption or radiolabeled substrate incorporation into macromolecules. The advantages of using fluorescent intracellular probes include the direct assessment of 1) physiological status; 2) specific metabolic activities; 3) gene expression; and 4) total cell densities. The ability to provide expeditious results pertaining to the presence and physiological activity of specific bacteria has direct application in a range of systems, including the monitoring of drinking water, wastewater, untreated and treated process waters (e.g., pharmaceuticals), and foods. Intracellular fluorescent probes have been used for assessing physiological activity after disinfection of biofilm-associated bacteria (Huang et al., 1995; Stewart et al. 1994; Yu and McFeters, 1994) and planktonic cells (Lisle et al., 1998, 1999).

2.6 Viral Pathogens

Enteroviruses are the cause of the most prevalent sewage-related waterborne diseases in developed countries (Cabelli, 1983). Enteroviruses are more persistent in treated wastewater effluent than coliform bacteria, which are commonly used to assess treatment efficacy (Englebrecht and Greenberg, 1978). Because of the high cost and intensive labor requirements of enumeration of enterovirus, bacterial viruses known as bacteriophage have been suggested as alternative indicators to coliforms for the assessment of water quality (Kott et al., 1974, 1978; Stetler, 1984). Coliphages, viruses that infect coliform bacteria such as *E. coli,* have been most commonly used in this context, although other bacteriophages have also been studied. Much of the justification for the study of coliphage behavior in nature has been to gain insight into the fate of human pathogenic enteric viruses. As a result, more is probably known about the ecology of coliphage than any other bacteriophage group.

Coliphage generally satisfy most of the criteria set forth for indicators of enteric viruses: they are generally present with enteric viruses; they are often present in the same numbers or higher numbers than enteric viruses; they are relatively resistant to water treatment and disinfection processes; they are a reasonable indicator of fecal or sewage pollution; they do not multiply in the environment; they are nonpathogenic to humans; and they are detectable by rapid, inexpensive methods.

Several drawbacks regarding the use of phages as models for enteric virus behavior have also been identified. These include the presence of enteric viruses in some water environments in the absence of coliphages (Borrego et al. 1990; Grabow et al., 1993); replication of certain somatic phages in water environments (Borrego et al., 1990; Grabow et al., 1984); excretion of human enteric viruses associated with waterborne diseases almost exclusively by humans (Grabow et al., 1995), while phage excretion can come from humans and animals (Grabow et al., 1993, 1995); variations between phages and enteric viruses in their efficiency of recovery from water. Because of these factors, phages should be considered members of the suite of indicator organisms and should not be considered to be absolute indicators of the presence of enteric viruses.

Coliphage MS-2 has been studied and applied extensively as a viral indicator in water and wastewater treatment applications. MS-2 is a single-stranded RNA coliphage that is classified into the family *Leviviridae* that consists of a simple capsid of cubic symmetry, 21 nm to 30 nm in diameter, containing single-stranded RNA as the genome. MS-2 coliphage infect and replicate in *E. coli* bacterial strains and are detected by plaque formation on bacterial laden semisolid agar media. Because of the similarities in overt physical characteristics to pathogenic enteroviruses, MS-2 has been used as a model virus in water treatment and disinfection studies (Havalaar et al., 1990; IAWPRC, 1991).

2.7 Case Study: Metropolitan Water Reclamation District of Greater Chicago

The Metropolitan Water Reclamation District of Greater Chicago (MWRDGC) operates several large facilities to treat municipal wastewater, including one of the largest municipal wastewater treatment facilities in the world, with a design capacity of 1200 million gallons/day (MGD). The MWRDGC has an interesting history in terms of technology, policy, and innovation, particularly as related to the issue wastewater effluent disinfection

Leadership within the MWRDGC has often challenged conventional thinking on topics relating to municipal wastewater treatment; in several cases, the approaches taken by the MWRDGC to solve water treatment and water quality problems have resulted in important innovations that have subsequently been adopted by other municipalities (e.g., MWRDGC's approach to disinfection, which is described in great detail in a series MWRDGC publications [Lue-Hing et al., 1976; Sedita et al., 1987 a,b]).

MWRDGC's history of disinfection includes several important events. In the late 1800s, a significant disease outbreak attributed to direct ingestion of untreated sewage occurred among Chicago residents. In turn, the consumption of contaminated water was attributed to the discharge of untreated municipal wastewater into Lake Michigan, close to the intake points for the public water supply. To remedy this situation, the direction of water flow in the Chicago River was reversed (away from Lake Michigan).

Since that time, wastewater treatment has improved dramatically. The MWRDGC operates facilities that produce effluent with low concentrations of total suspended solids (TSS) and BOD, relative to most secondary facilities. Beginning in July 1972, the MWRDGC implemented continuous chlorination of the effluent from all of its facilities. Soon thereafter, the district initiated testimony before the Illinois Pollution Control Board questioning the wisdom of chlorine-based disinfection. The MWRDGC began an extensive investigation of the advantages and disadvantages of chlorination, resulting in a series of reports. These reports illustrated a number of important drawbacks to the practice. In general terms, the MWRDGC demonstrated that water quality in the receiving waters downstream of their facilities, which at times contained as much as 90% effluent (i.e., very little dilution), was the same or better when chlorine-based disinfection was terminated than when chlorination was practiced. They concluded the following:

- Bacterial quality was improved by chlorination only in a short reach of receiving stream (roughly 10 to 15 river miles). Beyond this area, the concentrations of viable bacteria were similar with or without chlorination; in some cases, downstream bacterial counts were higher when chlorination was in effect.
- Among waterborne microorganisms in municipal wastewater, viruses represent the greatest threat to human health. However, chlorine is a poor virucide. Concentrations of viable enteric viruses in receiving streams were not significantly affected by chlorination. This conclusion was reached for a system that was based on a nitrified effluent. Therefore, the vast majority of residual chlorine present in the disinfection contact chamber would have been in the form of free chlorine. Free chlorine is generally regarded as a more effective antimicrobial agent than combined chlorine, particularly when viruses are the target organism.
- Chlorination resulted in a substantial reduction in fish populations in receiving streams. In addition, the absence of fish allowed for increased populations of nuisance insects in and around Chicago waterways. These water quality characteristics were attributed to the toxicological effects of chlorine on fish (and presumably other organisms) in the receiving streams.
- A related point of concern expressed by MWRDGC was the effect of chlorine on water chemistry. The receiving waters used by MWRDGC for effluent discharge become a resource for downstream communities. MWRDGC expressed concern that the large quantities of chlorinated DBPs generated by their facilities (e.g., chloroform generation was estimated to be roughly 4700 lb per day) was detrimental to downstream water users.

In summary, MWRDGC presented a compelling argument that chlorination did not yield the benefits that were normally ascribed to disinfection practices. The risks of microbial exposure were not substantially affected by chlorination. Moreover, MWRDGC argued that the practice of chlorination was actually detrimental to water quality, based on measurements of aquatic life and other water quality parameters. Again, it is important to note that these observations were made in receiving streams that at times were essentially undiluted municipal wastewater effluent from wastewater treatment facilities that employed fairly conventional, though effective, treatment operations.

Based on this evidence, the Illinois EPA granted MWRDGC's request to discontinue disinfection at its largest facilities (Stickney, Northside, Calumet, and Lemont water reclamation plants [WRPs]). Subsequent monitoring of effluents and receiving streams demonstrated improvements in water quality. Disinfection by chlorination/dechlorination continues at the three other district facilities (Egan, Kirie, and Hanover Park WRPs) on a seasonal basis.

CHAPTER 3.0

BACTERIAL COMMUNITY RESPONSES TO UV IRRADIATION AND CHLORINATION/DECHLORINATION

3.1 Background

Experiments were conducted to assess the long-term effects of UV irradiation and chlorination/dechlorination on indigenous bacterial communities. These experiments were designed to provide information regarding the effects of disinfectant exposure on bacteria at time scales well beyond those represented by conventional methods, where disinfected effluent samples are collected and assayed for viable indicator bacteria immediately after treatment.

3.2 Chemical Preparation

All solutions were prepared from reagent grade chemicals. Nanopure water was used in the preparation of all solutions and in sample dilution. Stock solutions of free chlorine were prepared from sodium hypochlorite (NaOCl) solution with 4% active chlorine (w/v) and diluted to 4.2 mM, then stored in aluminum foil-covered, glass-stoppered flasks. Concentrations of the stock solutions were determined titrimetrically before use.

3.3 Methods

3.3.1 Long-Term Effects of Disinfection

After disinfection, some sub-lethally damaged organisms may be able to repair disinfectant-induced damage. Together with organisms that retain viability following disinfection, it is then possible for the microbial community to regrow. An objective of this study was to investigate the extent of bacterial regrowth for different types of wastewater effluent after chlorination/dechlorination or UV irradiation. After treatment (i.e., batch disinfection or control), samples were incubated at constant temperature for six days in a respirometer to study the long-term respirometric behavior of the treated samples. Concentrations of viable FC bacteria, as well as total bacterial concentrations were used to assess microbial activity and community change. These measurements were conducted daily. During incubation, oxygen consumption in the treated effluent samples was recorded in real time using a respirometer.

3.3.1.1 Artificial Substrate

Substrate augmentation was achieved by adding acetic acid to the sample so that the final concentration of acetic acid in the sample was 14.1 mg/L (~15 mg/L of BOD). Substrate was added at this concentration to mimic the BOD concentration of typical receiving waters.

Acetic acid was selected as the substrate because it generally has a relatively low growth yield factor ($Y_{X/S}$), as compared with other common substrates, so it may have relatively little impact on the composition of microbial community (Shuler and Kargi, 1992a).

Another concern regarding the selection of the artificial substrate was the degree of reduction of the substrate. The degree of reduction (γ) for organic compounds may be defined as the number of equivalents of available electrons per gram atom carbon. A high degree of reduction indicates a low degree of oxidation. Acetic acid has moderate degree of reduction, which was deemed suitable for these experiments (Shuler and Kargi, 1992b).

CCA stock solution of artificial substrate was diluted from acetic acid (reagent grade) to a final concentration of 6328 mg/L before each experiment. A small volume (1.0 mL) of substrate stock solution was added into each 450 mL of sample to achieve a final substrate concentration of 14.06 mg/L.

3.3.1.2 Bacterial Concentration Quantification

Viable FC bacteria were quantified using method 9222D (*Standard Methods*, 1995). Samples were appropriately diluted and filtered using a 0.45 μm membrane (MF-Millipore Membrane). Filters were transferred to a sterile nutrient pad that had been pre-treated with m-FC broth, then incubated at 44.5°C for 24 hours. The m-FC broth was made with rosolic acid to reduce interference by other microorganisms. After incubation, colony-forming units (CFU) on each petri dish were recorded. These colonies were further classified as FC and non-FC based on colony color and appearance. Dilution water and m-FC broth were autoclaved before each use.

Total bacteria counts were performed using acridine orange staining and an epifluorescent microscope (*Standard Methods*, 1995). Samples used for total bacteria assays were preserved by addition of formalin (37% formaldehyde; 30 μL/mL sample) and stored in the dark at 4°C before staining. FC and total bacteria concentration were monitored every 24 hours during the 144-hour (six-day) incubation period.

Acridine orange stock solution was prepared at a w/v ratio of 0.1% in phosphate buffer (pH, 7.2). After mixing for 15 minutes, the solution was filtered using a 0.22-μm membrane to exclude possible interfering particles. The resulting solution was autoclaved after filtration.

3.3.1.3 Ammonium-N Concentration

Ammonium-N concentration was measured using an ammonia-selective electrode. This electrode was designed to measure ammonia instead of ammonium. However, the pH values in wastewater effluents are generally close to neutral. Under these conditions, most ammoniacal-N will be present in the protonated form (ammonium) rather than neutral form (ammonia). To measure ammonium-N concentration with this electrode, ammonium was converted to ammonia by increasing the pH.

Probe calibration was performed by analyses of a series of standards, which were prepared from a stock solution (1000 mg/L as N) of ammonium chloride (NH_4Cl). The stock solution was then stored in a glass-stoppered flask and diluted to the designated concentrations (0.1, 1, 10 mg/L as N) before each use. Samples and standards were equilibrated to room temperature before ammonia measurement. During the measurement of ammonia concentration, the pH was increased to >13 to facilitate detection. Adjustment of pH was achieved by addition of 2 mL of ionic strength adjustor (ISA) solution in 100 mL of sample under mild stirring. Ammonium-N concentration was estimated by comparison of the voltage signal from a sample with voltage signals developed from standard solutions. Membrane and electrode filling solution were replaced on a weekly basis; the electrode was stored in DI water when not in use.

3.3.1.4 Respirometer

A respirometer (OO-104 system, N-CON Systems Co., Inc., Crawford, Georgia) was used to measure oxygen consumption as a function of time in treated wastewater effluent samples. Each sample was placed in a water-bath incubator at 25°C under dark conditions with magnetic stirring. Each sample was capped with an O-ring to prevent gas leakage and linked to a gas valve that provided pure oxygen. An alkaline solution was prepared to absorb the carbon dioxide (CO_2) gas produced from the test sample during incubation. The alkaline solution was prepared by addition of five potassium hydroxide (KOH) pellets in 3 mL of nanopure water. The absorption of CO_2 caused a pressure decrease in the headspace; this resulted in release of an O_2 pulse to restore headspace pressure. Oxygen was released at a rate of ~0.17 mL O_2/pulse under 25°C and 1.0 atm.

3.3.1.5 Experimental Procedures

3.3.1.5.1 Sample Handling

Upon receipt, wastewater effluent samples were stored in six 1-L containers, then further split. To ensure sample homogeneity, four of the 1-L containers were poured into a sterile 4-L container containing a magnetic stirring bar. The remaining two 1-L containers were stirred at room temperature until the sample equilibrated to room temperature (22 °C to 25°C). The 2-L sample was then used in determination of ammonium-N concentration, DPD (N,N-diethly-p-phenylenediamine)/FAS titration, MIMS experiments, UV transmittance characterization, and dose-response tests of chlorination/dechlorination and UV irradiation.

After stirring for approximately 90 minutes, samples were split equally from the original container to four sterile, 1000-mL beakers containing a stirring bar for different treatments and subsequent assays. Preliminary chemical analysis of effluent sample was conducted before each treatment. Ammonium concentration was determined using the ammonia-selective probe described above. The absorbance behavior of the effluent sample was quantified using a UV/VIS spectrophotometer (Perkin Elmer, Lambda 20 UV/VIS spectrometer) over the wavelength range of 200 nm to 400 nm.

3.3.1.5.2 Chlorination/Dechlorination

Effluent samples were disinfected in bench-scale reactors under conditions that allowed precise, quantifiable disinfectant exposure histories. Disinfectant exposure scenarios were designed to allow effluent samples to comply with relevant discharge regulations at each facility, based on the concentrations of viable FC and residual chlorine. Preliminary dose-response experiments were conducted with effluent samples from each facility. The results of these experiments indicated that chlorination at 2.0 mg/L (as Cl_2), followed by dechlorination after 40 minutes of exposure yielded acceptable effluent quality. Graphical representations of preliminary dose-response experiments, as well as additional experiments that were conducted during the course of the research are provided in Appendix A.

After samples were equilibrated to room temperature, sample temperature and pH were recorded. A predetermined volume of stock free chlorine, prepared from sodium hypochlorite, was added to 1.0 L of sample to provide an initial concentration of 2.0 mg/L as Cl_2 in the sample. 150 mL to 200 mL of subsample was taken at 20 and 40 minutes after the addition of free chlorine; of this, 100 mL was used in DPD titration to determine concentrations of free and combined chlorine. The remaining sample was used for membrane filtration after addition of two small pellets of sodium thiosulfate for dechlorination. Each sample was appropriately diluted, if needed, and processed for the recovery or enumeration of bacteria using the protocols described above.

Another set of chlorination tests was set up following a similar procedure as described above for the respiration assay. 1.0 L of sample was chlorinated with an initial chlorine concentration of 2.0 mg/L (as Cl_2) for 1 hour and then dechlorinated with 6 mL sodium thiosulfate (1000 mg/L) solution. After dechlorination, artificial substrate (acetic acid) was added into the final solution before incubation for respiration test.

3.3.1.5.3 UV Irradiation

As with the batch chlorination procedure described above, it was necessary to define a UV exposure protocol to be used for bench-scale disinfection of effluent samples. Preliminary experiments demonstrated that a UV dose of 20 mJ/cm^2 allowed consistent compliance with effluent discharge regulations (see Appendix A). UV dose delivery was accomplished in reactor systems that allowed an essentially uniform, quantifiable UV dose to be delivered to effluent samples. Uniform dose delivery is beneficial in the interpretation of microbial responses to UV exposure. In general, continuous-flow UV disinfection systems deliver a broad distribution of UV doses, and as such, the interpretation of microbial responses to these exposures can be complicated and potentially misleading. A discussion of the reactor systems used to deliver these doses is provided below.

Prior to UV irradiation, the transmittance of an effluent sample at 254 nm was measured by a spectrophotometer. A collimated beam apparatus was used for the UV disinfection experiment. An 8-mL sample was poured into a petri dish with a diameter of 4.7 cm and exposed to the collimated UV beam. Samples were subjected to UV irradiation for exposure periods that were calculated to achieve doses of 5 mJ/cm^2, 10 mJ/cm^2, and 20 mJ/cm^2. Samples were mixed using a magnetic stirrer during UV exposure. Due to the limited size of the sample that could be effectively exposed in the collimated beam, multiple exposure events were necessary to accumulate sufficient volume at each dose to complete most of the microbiological assays. The exposed sample was then poured into a sterile flask that was covered with aluminum foil. This process was repeated until sufficient volume (8 mL to 100 mL) had been collected to complete membrane filtration assays. Transmittance of original and filtered samples was characterized using a UV-Visible spectrophotometer (Perkin Elmer, Lambda 20 UV/VIS spectrometer) before UV irradiation. For respiration experiments, in which large sample volumes (~1 L) were required, UV irradiation was achieved by using a capillary reactor to reduce sample collection time (see Chapter 4.0).

3.3.1.5.4 Respiration

As described previously, samples were split into four subsamples for bench-scale disinfectant exposure. Undisinfected effluent samples were collected from existing full-scale wastewater treatment facilities. Samples were kept on ice and shipped to allow analysis beginning the following day. The four treatments used in these experiments included (labels used to identify treatments are indicated in parentheses):

1) Original effluent without (w/o) addition of substrate (designated as Ori w/o)
2) Original effluent with (w) addition of substrate (designated as Ori w)
3) Chlorinated/dechlorinated effluent with addition of substrate (initial chlorine, 2.0 mg/L as Cl_2; contact time, 1.0 hour; dechlorinated with slight stoichiometric excess of sodium thiosulfate) (designated as Cl)
4) UV irradiated effluent with addition of substrate (UV dose, 20 mJ/cm^2) (designated as UV).

These four treatments were subjected to a respiration test for 144 hours using the respirometer system described previously. Consumption of oxygen was quantified automatically every 0.2 hours by the instrument, and measurements were recorded on a computer. FC and total bacteria concentration in each sample were monitored every 24 hours during the incubation period.

3.4 Results and Discussion

3.4.1 Long-Term Effect of Wastewater Disinfection

3.4.1.1 Characteristics of Effluent Samples

The focus was on characterization of the effects of different treatments on the quality of effluent samples collected from four municipal wastewater treatment facilities. All four facilities employ chlorine-based disinfection. Facility A employs primary clarification, activated sludge (with nitrification), and seasonal chlorination/dechlorination. The effluent from this facility is consistently of high quality in terms of conventional bulk parameters such as BOD_5 (< 5 mg/L, typ.) and TSS (< 5 mg/L, typ.). At the time of this research, Facility B employed conventional activated sludge treatment without nitrification and with seasonal chlorination/dechlorination. Effluent quality for samples collected during this research was typically poorer than from the other facilities, in terms of BOD_5, TSS, and ammonia concentration. Facility C employs nitrification and denitrification prior to discharge. Facility D, which subjects the effluent to conventional secondary treatment (without nitrification) and filtration, produces a high quality effluent in terms of BOD_5 and TSS, but with a relatively high concentration of NH_3-N. Facilities C and D both practice year-round chlorination/dechlorination and discharge their treated effluents to a marine environment.

To characterize the effects of wastewater disinfection on human health from the standpoint of human exposure to microorganisms, a number of possible scenarios must be considered. In this portion of the study, the inactivation and regrowth behavior of bacterial populations following disinfectant exposure were characterized. As a result, it is necessary to consider the total bacterial population, as well as the fraction of the bacterial population that expresses pathogenicity toward humans. Based on this logic, nine possible outcomes are possible (Figure 3-1). It is evident that not all disinfection scenarios should (necessarily) be considered effective in terms of reducing human exposure to pathogenic microorganisms. Figure 3-1 provides a schematic representation of the fate of bacteria subjected to disinfectant exposure. In the cases of viral and protozoan organisms, the circumstances may be somewhat different in that inactivation, as well as repair/recovery responses are likely to be substantially different than with bacteria.

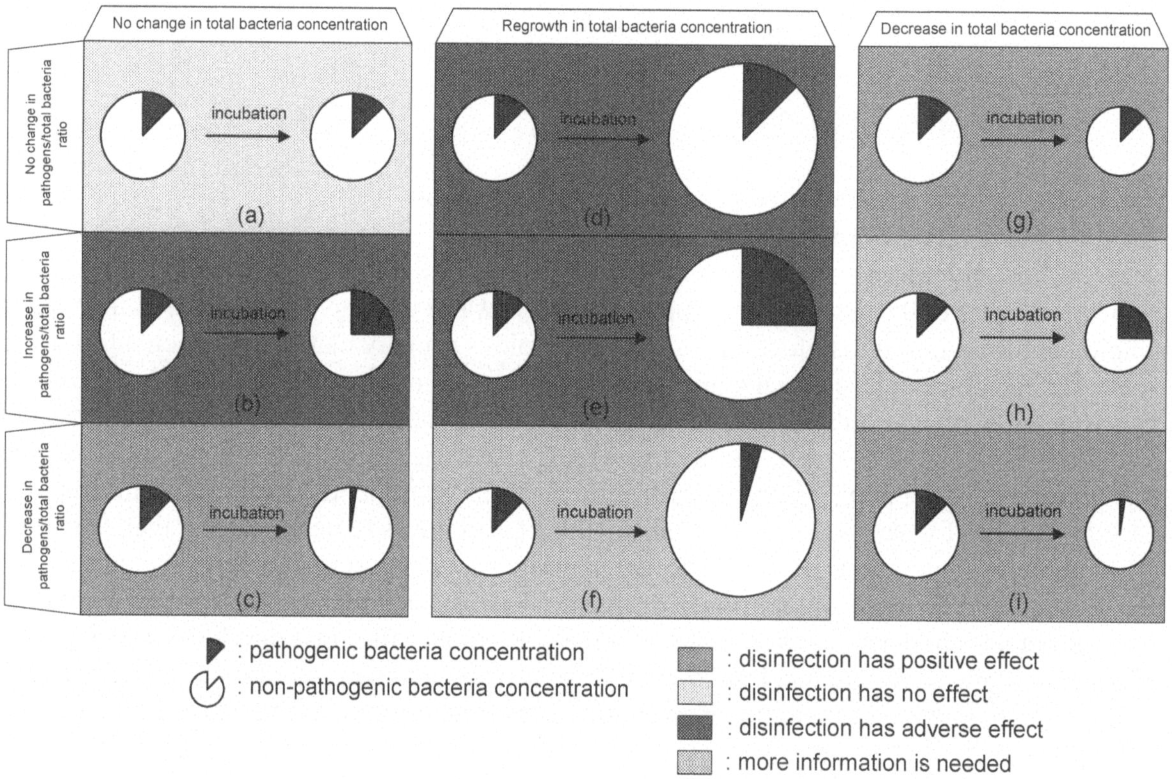

Figure 3-1. Conceptual Representation of the Possible Fates of Bacteria Following Disinfectant Exposure. Disinfection is considered to be antibacterially "effective" when the risk of human exposure to bacteria is reduced. Moving from left to right, the columns represent circumstances of no regrowth, regrowth, and decline in the total bacterial population, respectively. Moving from top to bottom, the rows represent circumstances in which the fraction of the bacterial population comprised of pathogenic bacteria does not change, increases, and decreases, respectively. Together, these two attributes (regrowth of the total bacterial population and changes in the fraction of pathogenic bacteria) determine the effectiveness of disinfection relative to human exposure to bacteria.

Figure 3-1 lists nine possible scenarios that could develop following disinfection. From this figure, one can judge the effectiveness of a disinfection process by variations in the total bacterial community, and the pathogenic fraction. For example, cases (c), (g) and (i) may represent a positive effect of disinfection since they imply a reduction in pathogenic bacteria; cases (a), (b), (d) and (e) have an adverse effect since pathogenic bacterial concentrations are not reduced. In cases (f) and (h), it is difficult to judge disinfection efficacy as judgment of antibacterial efficacy requires additional information, such as the concentration of pathogenic bacteria or indicator microorganisms. Cases for which disinfection is effective in reducing pathogenic microorganisms are indicated by green. Cases for which disinfection is not effective against pathogenic bacteria are indicated by red. Cases for which disinfection efficacy is not clear are indicated by gray.

To answer the question "is a disinfection process effective?" from the standpoint of bacterial risk, it is necessary to consider regrowth and the pathogen ratio. This involves investigating the impacts of upstream treatment processes, disinfection, and receiving waters on regrowth and the pathogen ratio. Under conditions of abundant substrate supply, rapid-growing microorganisms usually dominate populations. This is true in municipal wastewater treatment facilities, where the abundance of available organic substrates favors the growth of rapidly dividing bacteria, such as coliforms and pseudomonads. These dominant microbial populations in sewage, which gain a competitive advantage because of their high intrinsic growth rates, are rapidly displaced in competition with other microbial populations of receiving waters as the concentration of organic compounds diminishes, owing to decomposition and dilution; under lower nutrient conditions a more diverse community of slowly growing bacteria is favored.

For interpretation of the results of these experiments, several assumptions have been made, including:

- FC bacteria can be used as an indicator microorganism for pathogenic bacteria in disinfected wastewater effluent. This implicitly suggests that FC can be characterized as having similar a) susceptibility to disinfection processes; b) intrinsic growth rates; and c) requirements for nutrients as pathogenic bacteria.
- The substrate (acetic acid) used in this study can represent the substrate condition of receiving waters.
- The addition of substrate does not affect microbial ecology relationships during incubation.

It is clear that these assumptions are not entirely valid, because:

- Susceptibility to disinfection processes, intrinsic growth rates, and nutrient requirements for FC and pathogenic bacteria will be different; they are different even between two pathogenic bacteria. Coliform bacteria are commonly used as an "indicator" of microbial quality; however, no single species can truly represent the broad range of microbial pathogens that could be present in a municipal wastewater effluent.
- The nutrient conditions of receiving waters are site-specific, so it is impossible to find a substrate that can be representative of all situations.
- Acetic acid can be biodegraded easily; this will benefit bacteria with high intrinsic growth rates. The composition of easily biodegradable compounds in receiving waters will vary.

The conditions used in these experiments provided a common basis for examination of the behavior of municipal wastewater effluent from several facilities. It was possible to compare the long-term behavior of multiple samples collected from each of the facilities. The conditions of these experiments were believed to be representative of actual conditions in receiving waters; however, it is not reasonable to expect direct translation of these results to conditions in the respective receiving waters.

Assessments of disinfection efficacy have traditionally been based on the inactivation or removal of fecal indicators such as total coliform, FC, and fecal streptococci. However, there is little information about correlation between these indicator organisms and real pathogens, particularly in terms of long-term behavior. Although the assumptions listed above are not entirely justified, it is necessary to use this approach because relatively little information has been obtained regarding the ecological relationships between FC and pathogenic bacteria (many are not culturable). FC has been chosen as the target microorganism because it represents a common indicator microorganism for wastewater effluent regulation.

3.4.2 Statistical Analysis

Figure 3-2 depicts the design of the respiration experiments using wastewater effluents from the treatment facilities. Effluent samples from Facilities B and D were characterized by relatively high ammonium concentrations since the wastewater was not subjected to nitrification. Effluent samples from Facilities A and C were characterized by relatively low ammonium concentration because of the use of biological nitrification.

For each treatment facility, four effluent samples were collected on different dates and subjected to the same treatments. For each sample, four treatments (defined previously) were applied.

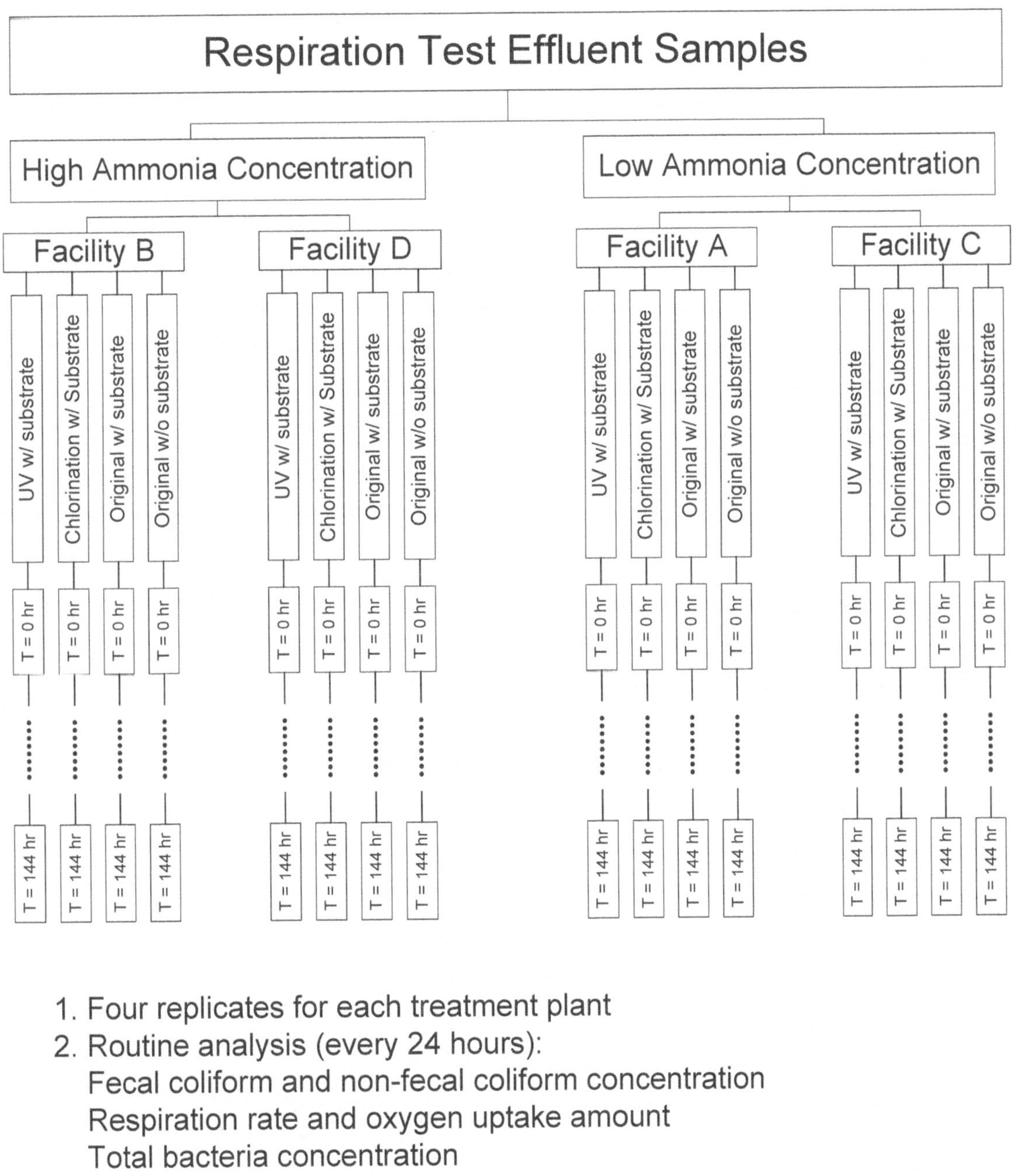

Figure 3-2. Illustration of Experimental Design Used in the Long-term Respiration Assays.
Four facilities were involved; four replicates (effluent collected on different date) were examined for each facility. Four treatments (Ori w, Ori w/o, Cl, UV) were applied to each replicate, and routine analyses were performed daily.

Disinfectant exposures were conducted in reactors that allowed controlled, quantifiable exposure; disinfectant exposures were selected to allow the treated samples to comply with relevant effluent discharge regulations, so as to provide relevant disinfectant exposures. After treatment, samples were incubated under constant temperature (25°C) for six days and routine analyses were performed every 24 hours. These analyses included: viable FC bacteria, non-FC, and total bacteria concentration. Results were used as input variables in the statistical analysis.

Statistical analyses were performed using the Statistical Analysis Software (SAS) system (Version 8). Since there were four treatment facilities, and each facility had four replicates, and each replicate involved four different treatments, the sample size was 64. Because the four facilities were not randomly selected, they were regarded as fixed factors; replicates from each facility were regarded as random factors. Hypothesis testing was conducted under $\alpha=0.05$, unless otherwise indicated.

Two levels of statistical analysis were involved: the first level of analysis was based on all 64 samples; and the second level of analysis was based on 16 samples from each facility. In the first level, the purpose was to investigate the effect of the different facilities on wastewater effluent quality after treatment. Prior to statistical model setup, a preliminary analysis of variance (ANOVA) test was performed to test the significance of each factor and interactions on output variables. The preliminary results showed that the effect of each individual factor (facility, replicate, and treatment) was significant; the interactions of facility and replicate were also significant. Therefore, it was inappropriate to make comparisons without considering the diversity of these factors. To account for these factors, each significant factor was included in the statistical model; replicates and facilities were used as blocks. In the second level of statistical analysis, the aim was limited to comparison of effect among different treatments on effluent sample in an individual facility. For each facility, comparisons were conducted based on 16 samples (4 treatments × 4 replicates / facility).

The output variables in this study included total oxygen uptake, viable FC concentration, total bacteria concentration, and the ratio of viable FC to total bacteria concentration. Because there were orders of magnitude differences in bacterial concentrations between disinfected and undisinfected samples, viable FC concentrations were $\log_{10}$ transformed.

3.4.2.1 Oxygen Uptake

Final oxygen consumption of each treatment was recorded after an incubation period of 144 hours. The resulting oxygen consumptions from each treatment were compared against initial ammonium concentrations based on 64 test samples. Results are shown in Figure 3-3. The superimposed straight line represents the theoretical nitrogenous oxygen demand with a slope of 4.3 g O_2/ g $[NH_4^+]$-N (Tchobanoglous and Burton, 1991). There was no clear trend between oxygen consumption and initial ammonium concentration when ammonium concentration was low (<1 mg/L), corresponding to effluent samples collected from nitrifying facilities. However, in samples collected from non-nitrifying facilities, where ammonium concentration was substantially higher, total oxygen consumption was strongly related to initial ammonium concentration, and higher than theoretical oxygen consumption based on ammonia oxidation was observed in most cases (except for UV irradiated samples). This suggests that most oxygen was consumed in the oxidation of ammonium (nitrification). For the range of ammonium concentration tested in this research ($[NH_3]_0 \approx 0–18$ mg/L as N), the non-disinfected samples, with or without substrate, usually yielded similar oxygen consumption, indicating that the artificial substrate did not cause a significant increase in overall oxygen uptake. Also, oxygen consumption in the non-disinfected samples was generally higher than in the disinfected samples.

Oxygen consumption of UV irradiated samples under high ammonium concentrations (14.35, 18.33 mg/L as N) was substantially lower than theoretical oxygen consumptions. The reason is not clear, but these two samples were both from Facility D, which uses filtration as an upstream treatment process. Guerrero and Jones (1996a) indicated that exposure to visible light (400 nm to 475 nm) and near-UV irradiation (300 nm to 375 nm) will cause inhibition to nitrite oxidizers and ammonium oxidizers. Furthermore, they found that ammonium oxidizers are more sensitive to photoirradiation than nitrite oxidizers. Guerrero and Jones (1996b) also investigated dark recovery of nitrifying bacteria after photoinhibition. They found that the recovery rates of *Nitrosomonas cryotolerans* and *Nitrosococcus oceanus* were slower when exposed to a short wavelength of radiation (300 nm), and the maximum recovery percentage appeared to be dependent on the wavelength of irradiation. However, no information regarding nitrifier response to UV near 254 nm was discussed in their studies, nor was the relative susceptibility of nitrifying bacteria to UV irradiation included. Therefore, similarities in observed responses of nitrifying bacteria in this research can only be made by inference of a similar set of processes being induced by solar UV radiation and the UV radiation that characterizes LP mercury (Hg) lamps, as used in this research.

The results of statistical analysis based on all 64 samples are summarized in Table 3-1. This analysis was conducted to compare the effect of different facilities on oxygen uptake based on all 64 samples. A p-value was determined to statistically evaluate the threshold curves. The null hypothesis (H_0) for the threshold curve was that the oxygen uptake values of the different facilities were statistically indistinguishable. The resulting p-value was <0.001, suggesting that the null hypothesis should be rejected. Duncan's multiple range test was performed to further investigate the difference between facilities, and the result showed that significant differences existed between high ammonium samples (Facilities B and D) and low ammonium samples (Facilities A and C).

Another analysis was performed to compare the effect of treatment on oxygen uptake. The resulting p-value of 0.065 implies no significant difference existed between treatments under an α (Type I error) value of 0.05. Duncan's multiple range test was performed to further investigate the difference between treatments, and the result showed a significant difference between treatment Ori w and treatment UV under α=0.05. Treatment Ori w/o and Cl was categorized in the same group; however, they were not significantly different from either treatment Ori w or treatment UV.

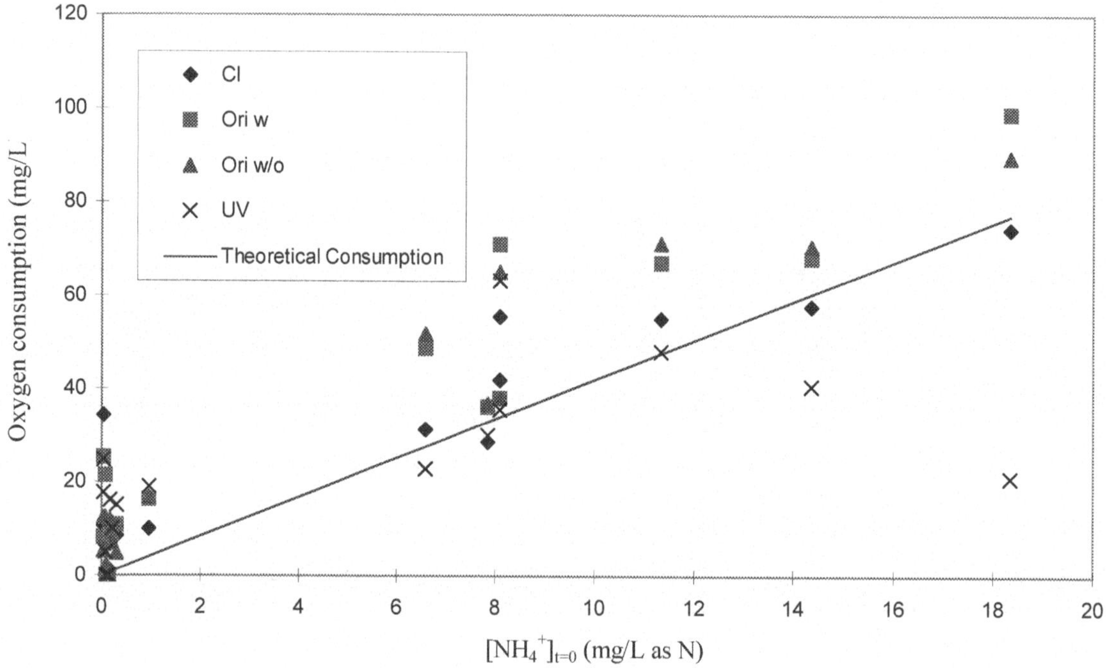

Figure 3-3. Results of Oxygen Uptake of Different Treatments at the End of 144-hour Incubation Period.
This figure includes results from all four facilities; the straight line represents the theoretical oxygen consumption based on ammonium-based (nitrogenous) BOD.

Table 3-1. Statistical Analyses on Total Oxygen Uptake (mg/L) at the End of Incubation Period.

Number in parenthesis represents mean value of each facility or treatment. Letter in parenthesis represents group classification of each facility or treatment.

Factor	p-value[1]	Mean value and Duncan's multiple range test
Facility	<0.001	Facility D (50.356), Facility B (44.536), Facility A (11.263), Facility C (7.619)
		Facility D (A), Facility B (A), Facility A (B), Facility C (B)
Treatment	0.065	Ori w (33.00), Ori w/o (31.00), Cl (26.679), UV (23.094)
		Ori w (A), Ori w/o (AB), Cl (AB), UV (B)

1. p-value of treatment

Table 3-2. Statistical Analyses of Each Facility on Total Oxygen Uptake at the End of Incubation Period.

Number in parenthesis represents mean value of each facility or treatment. Letter in parenthesis represents group classification of each facility or treatment.

Facility	p-value[1]	Mean value and Duncan's multiple range test
Facility B	0.0319	Ori w/o (51.2), Ori w (50.7), UV (38.275), Cl (37.968)
		Ori w/o (A), Ori w (A), UV (B), Cl (B)
Facility D	0.1019	Ori w (60.3), Ori w/o (58.73), Cl (50.6), UV (31.8)
		Ori w (A), Ori w/o (A), Cl (AB), UV (B)
Facility A	0.8769	Ori w (13.775), UV (11.950), Cl (11.225), Ori w/o (8.100)
		Ori w (A), UV (A), Cl (A), Ori w/o (A)
Facility C	0.2188	UV (10.35), Ori w (7.225), Cl (6.925), Ori w/o (5.975)
		UV (A), Ori w (A), Cl (A), Ori w/o (A)

1. p-value of treatment

Statistical analysis of each facility was conducted (Table 3-2). For non-nitrified effluents, which were characterized by relatively high ammonium concentration (Facilities B and D), disinfected samples (either UV or Cl) showed lower oxygen uptake than undisinfected samples (either Ori w or Ori w/o). However, the difference in oxygen uptake between two disinfection treatments was not statistically significant. The reduction on aerobic respiratory activity among the disinfected samples containing high ammonia-N was probably attributable to effects of the disinfectants on nitrifying bacteria. For Facility B, oxygen uptake was significantly different between disinfected sample and undisinfected sample, showing that disinfection may lead to diminished oxygen uptake. In effluent samples from facility D, the oxygen uptake of treatment UV was significantly lower than the undisinfected sample. The difference in oxygen uptake between the two non-disinfection treatments was not statistically significant, showing that addition of substrate may have no effect on effluent with high ammonium concentrations. For effluent containing relatively low ammonium concentration, treatment with Ori w/o yielded the lowest oxygen uptake. However, there was no clear difference among four treatments, suggesting that disinfection processes (UV or Cl) may not be able to lower the oxygen uptake effectively under this condition.

From these statistical analyses, several conclusions can be drawn:

1) For non-nitrified wastewater effluents containing high ammonium concentration, disinfection processes (chlorination/dechlorination or UV) appear to reduce subsequent oxygen consumption, probably as a result of disinfectant-induced stress on nitrifying bacteria. UV irradiation resulted in substantially greater suppression of nitrifier activity than chlorination/dechlorination.
2) For wastewater effluents containing low ammonium concentration, disinfection processes do not reduce oxygen consumption significantly.
3) Addition of substrate did not cause significant change in long-term oxygen consumption.

3.4.2.2 Fecal Coliform Concentration

FC bacteria were used as a bacterial indicator in this study. Their use for this purpose was motivated, in part, by extensive application in municipal wastewater effluent discharge regulations. Viable FC concentration in each sample was monitored every day during incubation. FC concentrations at t=120 hour were used for statistical analysis. Because variations in viable FC concentration of several orders of magnitude were observed, viable FC concentration used in statistical analysis was $\log_{10}$ transformed.

Table 3-3 summarizes the results of statistical analysis based on all 64 samples. The facility and the treatment factors were significant ($p < 0.001$). Non-nitrified effluents (Faciities B and D) showed higher FC concentration than the other two facilities after a five-day incubation period. After incubation for five days, viable FC concentration in disinfected samples was lower than that of non-disinfected samples. However, there was no significant difference in FC concentration between treatment Cl and treatment UV.

Table 3-4 summarizes results of statistical analysis of data for samples collected from each facility. Disinfected samples generally had lower viable FC concentration than non-disinfected samples after five days of incubation. However, it was difficult to differentiate between treatment Cl and treatment UV. In the non-disinfected samples of effluent from facility C, addition of artificial substrate resulted in a significant increase in FC concentration. This suggests that FC bacteria in effluent from facility C were influenced by addition of substrate. This phenomenon was only observed in Facility C samples.

Table 3-3. Statistical Analyses on log_{10} FC Concentration at t=120 Hour.
Number in parenthesis represents mean value of each facility or treatment. Letter in parenthesis represents group classification of each facility or treatment.

Factor	p-value[1]	Mean value and Duncan's multiple range test
Facility	<0.001	Facility B (3.10), Facility D (2.62), Facility A (1.99), Faclity C (0.98)
		Facility B (A), Facility D (A), Facility A (B), Facility C(C)
Treatment	<0.001	Ori w (3.06), Ori w/o (2.71), Cl (1.47), UV (1.46)
		Ori w (A), Ori w/o (A), Cl (B), UV (B)

1. p-value of treatment

Table 3-4. Statistical Analyses of Each Facility on log_{10} FC at t=120 Hour.
Number in parenthesis represents mean value of each facility or treatment. Letter in parenthesis represents group classification of each facility or treatment.

Facility	p-value[1]	Mean value and Duncan's multiple range test
Facility B	0.0252	Ori w/o (3.99), Ori w (3.85), Cl (3.09), UV (1.48)
		Ori w/o (A), Ori w (A), Cl (AB), UV (B)
Facility D	0.0182	Ori w (3.46), Ori w/o (2.97), UV (2.53), Cl (1.53)
		Ori w (A), Ori w/o (A), UV (AB), Cl (B)
Facility A	0.0089	Ori w/o (2.98), Ori w (2.84), UV (1.40), Cl (0.76)
		Ori w/o (A), Ori w (A), UV (B), Cl (B)
Facility C	0.0025	Ori w (2.07), Ori w/o (0.92), Cl (0.50), UV (0.44)
		Ori w (A), Ori w/o (B), Cl (B), UV (B)

1. p-value of treatment

3.4.2.3 Total Bacteria Concentration

Most microorganisms growing in nature have yet to be cultured in the laboratory (Pace, 1996). In fact, <1% of the microorganisms in natural water and soil samples are cultured in viable count procedures. Furthermore, some pathogenic bacteria are found to be viable nonculturable cells, such as *Legionella pneumophila*, *Salmonella enteritidis*, *Vibrio cholerae*, and *V. vulnificus* (Atlas and Bartha, 1998).

The bacterial populations in wastewater are diverse, and use of a single indicator microorganism to determine microbial quality is of limited value since it is likely to account for a very small fraction of the total bacteria present. Moreover, non-pathogenic bacteria may have different susceptibility to environmental changes than pathogenic bacteria. However, since it is not possible or practical to monitor the entire microbial community, examination of the behavior of a single organism, or class of organisms, may provide some insight into the effects of disinfection processes.

In addition to viable FC analyses, total bacteria concentration was monitored every day for each treatment using an acridine orange staining procedure and epifluorescence microscopy. Total bacteria concentrations at t=120 hour were used for statistical analysis. Table 3-5 summarizes the results of statistical analysis based on all 64 samples. The facility factor and treatment factor were significant ($p<0.001$). Effluents containing high ammonium concentration (Facilities B and D) showed relatively high concentrations of total bacteria after a five-day incubation period. The total bacteria concentration of the Facility B effluent was significantly higher than the other facilities. After incubation for five days, total bacteria concentration in chlorinated/dechlorinated samples was higher than samples from other treatments.

Table 3-5. Statistical Analyses on Total Bacteria Concentration at t=120 Hour.
Number in parenthesis represents mean value of each facility or treatment. Letter in parenthesis represents group classification of each facility or treatment.

Factor	p-value[1]	Mean value and Duncan's multiple range test
Facility	<0.001	Facility B (5.07E8), Facility D (1.81E8), Facility C (1.62E8), Facility A (7.63E7)
		Facility B (A), Facility D (B), Facility C (B), Facility A (B)
Treatment	<0.001	Cl (4.82E8), Ori w (2.01E8), UV (1.73E8), Ori w/o (1.60E8)
		Cl (A), Ori w (B), UV (B), Ori w/o (B)

1. p-value of treatment

Table 3-6. Statistical Analyses of Each Facility for Total Bacteria at t=120 Hour.
Number in parenthesis represents mean value of each facility or treatment. Letter in parenthesis represents group classification of each facility or treatment.

Facility	p-value[1]	Mean value and Duncan's multiple range test
Facility B	0.1733	Cl (6.89E8), Ori w (5.12E8), Ori w/o (4.33E8), UV (3.95E8)
		Cl (A), Ori w (A), Ori w/o (A), UV (A)
Facility D	0.0017	Cl (4.65E8), Ori w (1.1E8), UV (9.67E7), Ori w/o (5.45E7)
		Cl (A), Ori w (B), UV (B), Ori w/o (B)
Facility A	0.1516	Cl (2.00E8), UV (3.99E7), Ori w (3.59E7), Ori w/o (2.96E7)
		Cl (A), UV (A), Ori w (A), Ori w/o (A)
Facility C	0.0009	Cl (4.33E8), UV (9.23E7), Ori w (6.44E7), Ori w/o (5.84E7)
		Cl (A), UV (B), Ori w (B), Ori w/o (B)

1. p-value of treatment

Table 3-6 summarizes statistical analysis of data collected for samples from each facility. For samples collected at the end of the incubation period, samples from treatment Cl generally had the highest total bacteria concentration. In the effluent samples from Facilities C and D, total bacteria concentration in treatment Cl was significantly higher than samples from other treatments. In other facilities, the difference in total bacteria concentration was not statistically significant among the four treatments. The disinfection processes (chlorination/dechlorination or UV) can effectively reduce FC concentration; however, they have little or adverse long-term effect in terms of total bacteria. This may suggest FCs are more susceptible to chlorine and UV than other bacteria. Their use as an indicator may result in overestimation of disinfection efficacy.

3.4.2.4 Fecal Coliform/Total Bacteria Ratio

Since FC was used as the indicator microorganism in this study, the ratio of FC to total bacteria concentration can be viewed as an index of disinfection efficacy (Figure 3-1). Table 3-7 summarizes the results of statistical analysis based on FC/total bacteria ratio for all 64 samples. Neither the facility nor the treatment factor was significant (α level of 0.05).

Table 3-8 provides a summary of statistical analysis of data collected for samples from each facility. Fecal coliforms account for only a small percentage of total bacterial population in wastewater effluent regardless of treatment. Disinfected samples generally had lower FC/total bacteria ratios than non-disinfected samples at the conclusion of incubation. However, it was difficult to differentiate between treatments Cl and UV. As describe above, this may suggest that FC bacteria are more susceptible to conventional disinfection processes (chlorination/dechlorination or UV irradiation) than most other microorganisms. Note that in the non-disinfected samples from Facility C, addition of artificial substrate resulted in a significantly higher FC/total bacteria ratio. This may suggest that the FC percentage in Facility C effluent tends to be influenced by substrate addition. By comparing with total bacteria data of this facility, the increase of FC percentage may be attributable to both relatively high FC concentration and relatively low total bacteria concentration.

Table 3-7. Statistical Analyses on FC/Total Bacteria Concentration at t=120 Hour.
Number in parenthesis represents mean value of each facility or treatment. Letter in parenthesis represents group classification of each facility or treatment.

Factor	p-value[1]	Mean value and Duncan's multiple range test
Facility	0.1756	Facility D (7.39E-5), Facility B (1.47E-5), Facility A (5.97E-6), Facility C (5.8E-7)
		Facility D (A), Facility B (A), Facility A (A), Facility C (A)
Treatment	0.2943	Ori w (7.26E-5), Ori w/o (2.84E-5), UV (2.52E-6), Cl (1.90E-6)
		Ori w (A), Ori w/o (A), UV (A), Cl (A)

1. p-value of treatment

Table 3-8. Statistical Analyses of Each Facility for FC/Total Bacteria Concentration at t=120 Hour.
Number in parenthesis represents mean value of each facility or treatment. Letter in parenthesis represents group classification of each facility or treatment.

Facility	p-value[1]	Mean value and Duncan's multiple range test
Facility B	0.023	Ori w/o (3.17E-5), Ori w (2.00E-5), Cl (4.94E-6), UV (2.26E-6)
		Ori w/o (A), Ori w (AB), Cl (B), UV (B)
Facility D	0.4024	Ori w (2.29E-4), Ori w/o (5.84E-5), UV (6.5E-6), Cl (1.7E-6)
		Ori w (A), Ori w/o (A), UV (A), Cl (A)
Facility A	0.0141	Ori w/o (1.80E-5), Ori w (5.95E-6), Cl (0), UV (0)
		Ori w/o (A), Ori w (B), Cl (B), UV (B)
Facility C	< 0.0001	Ori w (1.94E-6), Ori w/o (2.69E-7), UV (5.02E-8), Cl (9.20E-9)
		Ori w (A), Ori w/o (B), UV (B), Cl (B)

1. p-value of treatment

3.4.3 Regrowth Potential

As described previously, an important factor affecting long-term disinfection efficacy is regrowth potential. Based on regrowth condition and FC (or indicator) to total bacteria ratio, the long-term outcomes of disinfection processes can be further divided into the nine scenarios illustrated in Figure 3-1. It is also possible to use these nine scenarios to classify disinfection processes as effective or ineffective, while providing information regarding microbial community change at the same time.

Based on the 16 experimental runs (four treatment facilities × four replicates / facility), four different treatments were applied to the wastewater sample, and FC and total bacteria concentrations were recorded from t=0 hour (immediately after treatment) to t=144 hour. The ratio of FC to total bacteria was then calculated for each sampling time. Since there were four replicates involved in each treatment facility, an average value of the four replicates was used for representation of the total bacteria concentration and fecal to total bacteria ratio. For each parameter (FC/total bacteria ratio, total bacteria concentration, FC concentration) at each incubation time, the data were normalized with respect to the initial values (t=0 hour) to facilitate the assessment procedure. Based on this information, classification of disinfection process proceeded using Figure 3-1.

Table 3-9 lists the assessment results of Facility B for four different treatments with time. In this table, the normalized value of ratio (fecal/total) was used to indicate the increase (normalized ratio >1) or decrease (normalized ratio <1) of the indicator percentage. If the normalized ratio was >1, the sample was described by the second row in Figure 3-1 and was thus categorized as "R2." Similarly, if the normalized ratio was <1, it could be categorized as "R3." Total bacteria concentrations were also normalized with respect to respective initial values at t=0 hour. If the normalized total bacteria concentration was >1, implying an increase in the total bacteria concentration, it was described by the second column in Figure 3-1 and was categorized as C2." By locating the position based on row and column in Figure 3-1, the long-term disinfection efficiency was assessed by case (a) to case (i). For most cases in Figure 3-1, the disinfection efficacy can be identified unambiguously. However, in cases (f) and (h), it is difficult to characterize disinfection efficacy based on these data. Viable FC concentration can provide additional information for these "gray zones"; an increase of FC was assigned a "+" symbol and a decrease of FC was assigned a "-" symbol.

For Facility B, UV irradiated samples changed from case (e) (red)→ (b) (red) → (h) (gray) → (i) (green), indicating that samples actually underwent regrowth initially (t = 24 – 48 hour) and then microbial quality improved with time after the regrowth phase when FC bacteria were used as the indicator. Further, the observation of cases (e) and (b) in UV irradiated samples implied that regrowth did occur at t=24 hour and t=48 hour, respectively. For "gray zones" at t=72 – 120 hr, FC data suggest that microbial quality may have actually been worse than the initial condition. For chlorinated/dechlorinated samples, they fell into case (e) at all times, meaning that not only total bacterial concentration increased, but also the FC percentage also increased, and thus deteriorated the microbial quality of disinfected wastewater. Non-disinfected samples (Ori w and Ori w/o) both changed from case (i) (green)→ (c) (green) → (f) (gray) → (i) (green) or (c) (green). For those samples (Ori w at 72 hour, 96 hour; Ori w/o at 96 hour) that fall into the case (f) (gray zone), the FC data indicates that the microbial quality of these samples actually did not change. Generally speaking, when FC bacteria were used as the indicator, disinfected samples became more "unstable" or had higher regrowth potential than undisinfected samples when compared with initial values (t=0 hour). However, disinfected samples generally had lower absolute FC concentrations than undisinfected samples.

Table 3-10 lists the assessment results for Facility D for four different treatments with time. For UV irradiated samples, they changed from case (e) (red) → (a) (yellow) → (h) (gray), indicating that regrowth occurred initially, followed by a decrease in FC concentration. Chlorinated/dechlorinated samples changed from case (e) (red) → (f) (gray) → (e) (red), suggesting that these samples were unstable when FC was used as the indicator. For non-disinfected samples, a similar trend was followed ((f) (gray) → (i) (green) or (c) (green)). The FC data suggested that non-disinfected samples were generally more stable than disinfected samples because they usually fell into green zones.

Table 3-11 lists the assessment results for Facility A for four treatments with time. The UV irradiated samples changed from case (f) (gray) → (i) (green). By comparing with FC data, no regrowth was observed for FC, and the occurrence of "gray zones" can be attributable to regrowth of total bacteria. Chlorinated/dechlorinated samples changed from case (f) (gray) → (e) (red), and gray zones were attributable to regrowth of total bacteria. For non-disinfected samples, they followed a similar trend (e) (red) or (f) (gray) → (i) (green). The FC data suggested that non-disinfected samples were generally more stable than disinfected samples because they fell into the green zones most of the time.

Table 3-12 lists the assessment results for Facility C for four treatments with time. The UV irradiated samples changed from case (e) (red) → (h) (gray). By comparing with FC data, regrowth occurred in FC and the occurrence of "gray zones" can be attributable to regrowth of FC and decrease of total bacteria. Chlorinated/ dechlorinated samples changed back and forth from case (f) (gray) → (e) (red), and gray zone can be attributable to zero fecal/total ratios since no FC was observed. Non-disinfected samples fell into case (i) (green) at all times, suggesting that these were generally more stable than disinfected samples.

Table 3-9. Assessment Result of Disinfection Efficiency of Wastewater Effluent from Facility B.

(a) UV disinfection

Time	0 hr	24 hr	48 hr	72 hr	96 hr	120 hr	144 hr
Ratio (Fecal/total)	1.93E-06	4.87E-05	2.71E-05	3.84E-06	2.39E-06	2.26E-06	4.10E-07
Normalized	1.00	25.24	14.03	1.99	1.24	1.17	0.21
Category	-	R2	R2	R2	R2	R2	R3
Total bacteria concentration	6.02E+08	6.99E+08	5.99E+08	4.33E+08	4.32E+08	3.95E+08	4.63E+08
Normalized	1.00	1.16	1.00	0.72	0.72	0.66	0.77
Category	-	C2	C1	C3	C3	C3	C3
FC concentration	495	10000	8500	2250	1850	2025	300
Normalized	1.00	20.20	17.17	4.55	3.74	4.09	0.61
Category	-	e	b	h	h	h	i
Additional information				+	+	+	

(b) Chlorination/dechlorination

Time	0 hr	24 hr	48 hr	72 hr	96 hr	120 hr	144 hr
Ratio (Fecal/total)	2.79E-06	1.09E-05	6.60E-06	6.11E-06	3.85E-06	4.94E-06	3.55E-06
Normalized	1.00	3.89	2.36	2.19	1.38	1.77	1.27
Category	-	R2	R2	R2	R2	R2	R2
Total bacteria concentration	5.22E+08	7.37E+08	7.83E+08	5.83E+08	5.93E+08	6.89E+08	5.99E+08
Normalized	1.00	1.41	1.50	1.12	1.14	1.32	1.15
Category	-	C2	C2	C2	C2	C2	C2
FC concentration	715	5975	11421.5	2732.5	1992.5	2475	1132.5
Normalized	1.00	8.36	15.97	3.82	2.79	3.46	1.58
Category	-	e	e	e	e	e	e
Additional information							

Table 3-9. (continued)

(c) Ori w

Time	0 hr	24 hr	48 hr	72 hr	96 hr	120 hr	144 hr
Ratio (Fecal/total)	8.69E-04	1.23E-04	4.01E-05	4.22E-05	3.35E-05	2.00E-05	1.31E-05
Normalized	1.00	0.14	0.05	0.05	0.04	0.02	0.02
Category	-	R3	R3	R3	R3	R3	R3
Total bacteria concentration	5.34E+08	4.16E+08	4.63E+08	5.40E+08	6.62E+08	5.12E+08	5.09E+08
Normalized	1.00	0.78	0.87	1.01	1.24	0.96	0.95
Category	-	C3	C3	C1, 2	C2	C1, 3	C1, 3
FC concentration	281000	48500	16250	18250	16125	7700	5825
Normalized	1.00	0.17	0.06	0.06	0.06	0.03	0.02
Category	-	i	i	c f	f	c, i	c, i
Additional information				-	-		

(d) Ori w/o

Time	0 hr	24 hr	48 hr	72 hr	96 hr	120 hr	144 hr
Ratio (Fecal/total)	5.57E-04	1.09E-04	7.00E-05	4.22E-05	2.50E-05	3.17E-05	4.03E-05
Normalized	1.00	0.19	0.13	0.08	0.04	0.06	0.07
Category	-	R3	R3	R3	R3	R3	R3
Total bacteria concentration	5.49E+08	3.69E+08	4.35E+08	5.55E+08	5.93E+08	4.33E+08	3.72E+08
Normalized	1.00	0.67	0.79	1.01	1.08	0.79	0.68
Category	-	C3	C3	C1	C2	C3	C3
FC concentration	215500	22000	21000	20250	12000	10825	7275
Normalized	1.00	0.10	0.10	0.09	0.06	0.05	0.03
Category	-	i	i	c	f	i	i
Additional information					-		

Table 3-10. Assessment Result of Disinfection Efficiency of Wastewater Effluent from Facility D.

(a) UV disinfection

Time	0 hr	24 hr	48 hr	72 hr	96 hr	120 hr	144 hr
Ratio (Fecal/total)	9.01E-06	1.43E-05	3.30E-05	4.47E-05	2.23E-05	6.52E-06	1.59E-05
Normalized	1.00	1.59	3.67	4.97	2.47	0.72	1.77
Category	-	R2	R2	R2	R2	R3	R2
Total bacteria concentration	9.44E+07	5.09E+08	3.79E+08	1.16E+08	1.26E+08	9.67E+07	7.90E+07
Normalized	1.00	5.39	4.01	1.22	1.33	1.02	0.84
Category	-	C2	C2	C2	C2	C1,2	C3
FC concentration	640.25	2625	6500	5250	3275	670	990
Normalized	1.00	4.10	10.15	8.20	5.12	1.05	1.55
Category	-	e	e	e	e	a	h
Additional information							+

(b) Chlorination/dechlorination

Time	0 hr	24 hr	48 hr	72 hr	96 hr	120 hr	144 hr
Ratio (Fecal/total)	1.18E-06	4.43E-05	1.56E-06	9.09E-06	8.18E-07	1.70E-06	3.59E-06
Normalized	1.00	37.74	1.33	7.74	0.70	1.45	3.05
Category	-	R2	R2	R2	R3	R2	R2
Total bacteria concentration	8.53E+07	3.21E+08	4.54E+08	4.27E+08	4.14E+08	4.65E+08	4.09E+08
Normalized	1.00	3.77	5.32	5.01	4.86	5.45	4.80
Category	-	C2	C2	C2	C2	C2	C2
FC concentration	61.5	7050	500	2625	350	902.5	2040
Normalized	1.00	114.63	8.13	42.68	5.69	14.67	33.17
Category	-	e	e	e	f	e	e
Additional information					+		

Table 3-10. (continued)

(c) Ori w

Time	0 hr	24 hr	48 hr	72 hr	96 hr	120 hr	144 hr
Ratio (Fecal/total)	1.54E-03	1.21E-03	9.41E-05	1.21E-04	1.29E-04	2.29E-04	1.25E-04
Normalized	1.00	0.78	0.06	0.08	0.08	0.15	0.08
Category	-	R3	R3	R3	R3	R3	R3
Total bacteria concentration	1.12E+08	1.76E+08	1.67E+08	1.21E+08	1.06E+08	1.10E+08	7.00E+07
Normalized	1.00	1.58	1.49	1.08	0.95	0.98	0.63
Category	-	C2	C2	C2	C3	C1,3	C3
FC concentration	237375	125625	16213.5	12962.5	9340	4541.25	2718
Normalized	1.00	0.53	0.07	0.05	0.04	0.02	0.01
Category	-	f	f	f	i	c, i	i
Additional information		-	-	-			

(d) Ori w/o

Time	0 hr	24 hr	48 hr	72 hr	96 hr	120 hr	144 hr
Ratio (Fecal/total)	3.00E-03	1.55E-03	3.45E-04	1.65E-04	3.80E-05	5.84E-05	4.41E-05
Normalized	1.00	0.52	0.12	0.05	0.01	0.02	0.01
Category	-	R3	R3	R3	R3	R3	R3
Total bacteria concentration	8.54E+07	1.25E+08	1.03E+08	8.17E+07	8.10E+07	5.45E+07	4.25E+07
Normalized	1.00	1.46	1.21	0.96	0.95	0.64	0.50
Category	-	C2	C2	C1,3	C1,3	C3	C3
FC concentration	194958.25	93675	18900	8887.5	3137.5	2265	1282
Normalized	1.00	0.48	0.10	0.05	0.02	0.01	0.01
Category	-	f	f	c, i	c, i	i	i
Additional information		-	-				

Table 3-11. Assessment Result of Disinfection Efficiency of Wastewater Effluent from Facility A.

(a) UV disinfection

Time	0 hr	24 hr	48 hr	72 hr	96 hr	120 hr	144 hr
Ratio (Fecal/total)	1.01E-06	0.00E+00	0.00E+00	0.00E+00	0.00E+00	0.00E+00	0.00E+00
Normalized	1.00	0.00	0.00	0.00	0.00	0.00	0.00
Category	-	R3	R3	R3	R3	R3	R3
Total bacteria concentration	5.63E+07	3.28E+08	5.89E+08	4.00E+08	1.03E+08	3.99E+07	3.69E+07
Normalized	1.00	5.83	10.46	7.11	1.82	0.71	0.65
Category	-	C2	C2	C2	C2	C3	C3
FC concentration	55	0	0	0	0	0	0
Normalized	1.00	0.00	0.00	0.00	0.00	0.00	0.00
Category	-	f	f	f	f	i	i
Additional information		-	-	-	-		

(b) Chlorination/dechlorination

Time	0 hr	24 hr	48 hr	72 hr	96 hr	120 hr	144 hr
Ratio (Fecal/total)	1.43E-07	0.00E+00	0.00E+00	0.00E+00	0.00E+00	0.00E+00	2.36E-06
Normalized	1.00	0.00	0.00	0.00	0.00	0.00	16.53
Category	-	R3	R3	R3	R3	R3	R2
Total bacteria concentration	6.31E+07	2.85E+08	4.06E+08	1.69E+08	1.72E+08	2.00E+08	2.16E+08
Normalized	1.00	4.51	6.43	2.68	2.73	3.17	3.43
Category	-	C2	C2	C2	C2	C2	C2
FC concentration	9	0	0	0	0	0	500
Normalized	1.00	0.00	0.00	0.00	0.00	0.00	55.56
Category	-	f	f	f	f	f	e
Additional information		-	-	-	-	-	

Table 3-11. (continued)

(c) Ori w

Time	0 hr	24 hr	48 hr	72 hr	96 hr	120 hr	144 hr
Ratio (Fecal/total)	1.53E-04	6.08E-04	0.00E+00	5.62E-05	6.11E-06	5.95E-06	3.30E-06
Normalized	1.00	3.97	0.00	0.37	0.04	0.04	0.02
Category	-	R2	R3	R3	R3	R3	R3
Total bacteria concentration	7.16E+07	6.32E+08	3.02E+07	2.44E+07	2.64E+07	3.59E+07	4.54E+07
Normalized	1.00	8.83	0.42	0.34	0.37	0.50	0.63
Category	-	C2	C3	C3	C3	C3	C3
FC concentration	9850	550000	0	1500	150	225	175
Normalized	1.00	55.84	0.00	0.15	0.02	0.02	0.02
Category	-	e	i	i	i	i	i
Additional information							

(d) Ori w/o

Time	0 hr	24 hr	48 hr	72 hr	96 hr	120 hr	144 hr
Ratio (Fecal/total)	1.34E-04	1.15E-04	4.96E-05	4.17E-05	3.61E-05	1.79E-05	1.00E-05
Normalized	1.00	0.86	0.37	0.31	0.27	0.13	0.07
Category	-	R3	R3	R3	R3	R3	R3
Total bacteria concentration	6.61E+07	9.01E+07	6.40E+07	2.97E+07	2.69E+07	2.96E+07	4.77E+07
Normalized	1.00	1.36	0.97	0.45	0.41	0.45	0.72
Category	-	C2	C3	C3	C3	C3	C3
FC concentration	9350	9850	2600	1350	970	505	475
Normalized	1.00	1.05	0.28	0.14	0.10	0.05	0.05
Category	-	f	i	i	i	i	i
Additional information		+					

Table 3-12. Assessment Result of Disinfection Efficiency of Wastewater Effluent Facility C.

(a) UV disinfection

Time	0 hr	24 hr	48 hr	72 hr	96 hr	120 hr	144 hr
Ratio (Fecal/total)	7.46E-09	1.97E-08	5.94E-08	4.06E-08	7.84E-08	5.02E-08	9.35E-09
Normalized	1.00	2.64	7.97	5.44	10.50	6.73	1.25
Category		R2	R2	R2	R2	R2	R2
Total bacteria concentration	2.02E+08	8.03E+08	3.93E+08	8.11E+07	1.03E+08	9.23E+07	1.62E+08
Normalized	1.00	3.97	1.94	0.40	0.51	0.46	0.80
Category	-	C2	C2	C3	C3	C3	C3
FC concentration	2	25	1.75	5.75	7.75	4.5	2
Normalized	1.00	12.50	0.88	2.88	3.88	2.25	1.00
Category	-	e	e	h	h	h	h
Additional information				+	+	+	-/+

(b) Chlorination/dechlorination

Time	0 hr	24 hr	48 hr	72 hr	96 hr	120 hr	144 hr
Ratio (Fecal/total)	7.35E-10	0.00E+00	3.60E-08	5.21E-09	0.00E+00	9.20E-09	0.00E+00
Normalized	1.00	0.00	49.00	7.09	0.00	12.52	0.00
Category	-	R3	R2	R2	R3	R2	R3
Total bacteria concentration	2.28E+08	5.34E+08	7.51E+08	4.73E+08	3.92E+08	4.33E+08	3.99E+08
Normalized	1.00	2.34	3.29	2.08	1.72	1.90	1.75
Category	-	C2	C2	C2	C2	C2	C2
FC concentration	0.25	0	25	1.75	0	5	0
Normalized	1.00	0.00	100.00	7.00	0.00	20.00	0.00
Category	-	f	e	e	f	e	f
Additional information		-			-		-

Table 3-12. (continued)

(c) Ori w

Time	0 hr	24 hr	48 hr	72 hr	96 hr	120 hr	144 hr
Ratio (Fecal/total)	7.28E-06	1.39E-06	3.91E-06	3.19E-06	2.06E-06	1.97E-06	1.07E-06
Normalized	1.00	0.19	0.54	0.44	0.28	0.27	0.15
Category	-	R3	R3	R3	R3	R3	R3
Total bacteria concentration	2.41E+08	2.20E+08	8.37E+07	6.04E+07	6.97E+07	6.44E+07	7.77E+07
Normalized	1.00	0.91	0.35	0.25	0.29	0.27	0.32
Category	-	C3	C3	C3	C3	C3	C3
FC concentration	1925	222.5	316.25	192.5	123	129	93
Normalized	1.00	0.12	0.16	0.10	0.06	0.07	0.05
Category	-	i	i	i	i	i	i
Additional information							

(d) Ori w/o

Time	0 hr	24 hr	48 hr	72 hr	96 hr	120 hr	144 hr
Ratio (Fecal/total)	1.05E-05	4.33E-06	3.18E-06	1.31E-06	5.94E-07	2.69E-07	1.76E-07
Normalized	1.00	0.41	0.30	0.12	0.06	0.03	0.02
Category	-	R3	R3	R3	R3	R3	R3
Total bacteria concentration	2.18E+08	9.81E+07	4.28E+07	7.09E+07	6.37E+07	5.84E+07	1.21E+08
Normalized	1.00	0.45	0.20	0.33	0.29	0.27	0.56
Category	-	C3	C3	C3	C3	C3	C3
FC concentration	2400	445	135	108	43	14.5	9.333333 333
Normalized	1.00	0.19	0.06	0.05	0.02	0.01	0.00
Category	-	i	i	i	i	i	i
Additional information							

3.5 Summary and Conclusions

The purpose of these experiments was to characterize the long-term effects of contemporary disinfection practices in bacterial populations present in municipal wastewater effluent. Several general conclusions can be drawn:

- Wastewater effluent tended to become less stable after disinfection (Chlorination/dechlorination or UV irradiation) than non-disinfected samples in terms of regrowth potential, based on the use of viable FC concentration as the indicator.
- Disinfection processes can result in reduced FC concentration, even after regrowth. The FC concentrations in disinfected samples were generally lower than non-disinfected samples.
- FC, when used as an indicator, may overestimate disinfection efficacy or microbial quality of disinfected samples, since they are relatively susceptible to common disinfectants (chlorine and UV) and they have higher die-off rate than other microorganisms. This study provides an example of how FC concentration changes with time in a microbial community after it is subjected to disinfection.
- Dark repair following UV irradiation may play an important role relative to the regrowth potential of UV disinfected samples. Similarly, "dark" (i.e., non-photochemical) repair mechanisms may also play a role in the fate of chlorinated samples.
- Addition of substrate was not found to increase regrowth potential significantly. This result may not be extrapolated to pathogenic bacteria since their susceptibility and regrowth potential may be substantially different than FC.
- Based on the long-term trends in FC and total bacterial concentrations (Tables 3-11 through 3-14), wastewater effluents respond more favorably to UV irradiation than to chlorination/dechlorination. For the effluent samples evaluated, bacterial quality (as defined by these two measures) was consistently better in UV irradiated samples than in samples that had been subjected to chlorination/dechlorination.
- In comparison with disinfected samples (by either chlorination/dechlorination or UV irradiation), and in comparison with samples prior to application of disinfection (Tables 3-11 through 3-14), non-disinfected samples consistently showed improved bacterial quality, as defined by the concentration of FC bacteria and total bacterial composition.

CHAPTER 4.0

CAPILLARY FLOW UV REACTOR

4.1 Background

Delivery of a well-defined, quantifiable, uniform dose of UV radiation is essential to the interpretation of microbial UV dose-response behavior. Among common reactor systems available to apply UV radiation to aqueous suspensions of microorganisms, important limitations exist. In laboratory (bench-scale) research with UV radiation, it is common to use a collimated beam in conjunction with a shallow, well-mixed batch reactor. These systems can be capable of satisfying the dosing requirements defined above; however, they are normally capable of irradiating only small volumes of liquid (generally <10 mL to 20 mL per batch). Under some circumstances, substantially larger volumes are required for analysis; when this is the case, the use of a collimated beam system can become impractical.

Continuous-flow UV systems can generate large volumes of treated water. However, in virtually all circumstances, continuous-flow reactors will deliver a broad distribution of UV doses. This complicates the interpretation of microbial (or chemical) assays from samples treated by continuous-flow systems, at least in terms of the definition of dose-response behavior. Moreover, this suggests that the interpretation of microbial UV "dose-response" behavior based on samples collected from conventional continuous-flow UV reactors should be viewed with considerable skepticism, unless the system has been designed and operated under conditions that will allow delivery of a well-defined, quantifiable dose.

Therefore, the need was perceived to develop a continuous-flow reactor system that is capable of delivering a uniform, quantifiable dose of UV radiation to comparatively large volumes of water. The development of this system was essential for the conduct of the long-term respirometry assays and should be useful in other applications.

4.2 Materials and Methods

4.2.1 Actinometer

A chemical actinometer was generated by mixing a solution of 0.6 M potassium iodide (KI) and 0.1 M potassium iodate in 0.01 M sodium borate/phosphate buffer (pH 9.25). The iodide/iodate actinometric solution absorbed germicidal UV radiation but little ambient light (Figure 4-1). Iodate acts as an electron scavenger and prevents the back reaction of the free electron with the iodine atom following UV excitation of KI. Irradiation results in the formation of triiodide (I_3^-), with I_3^- production rate being proportional to UV intensity. I_3^- concentration was quantified spectrophotometrically based on absorbance measurements at 352 nm. The actinometer system was also subjected to UV irradiation using a flat-plate collimated beam system. By performing these experiments over a broad range of UV doses (quantified by radiometry), it was possible to define the UV dose-response behavior of the actinometer system. These findings were consistent with previously published photochemical kinetic data for the iodide/iodate actinometer system (Rahn, 1993; 1997). UV dose delivery by the capillary-flow reactor was characterized by comparison of I_3^- formation in a capillary-flow reactor with the data from the collimated-beam experiments.

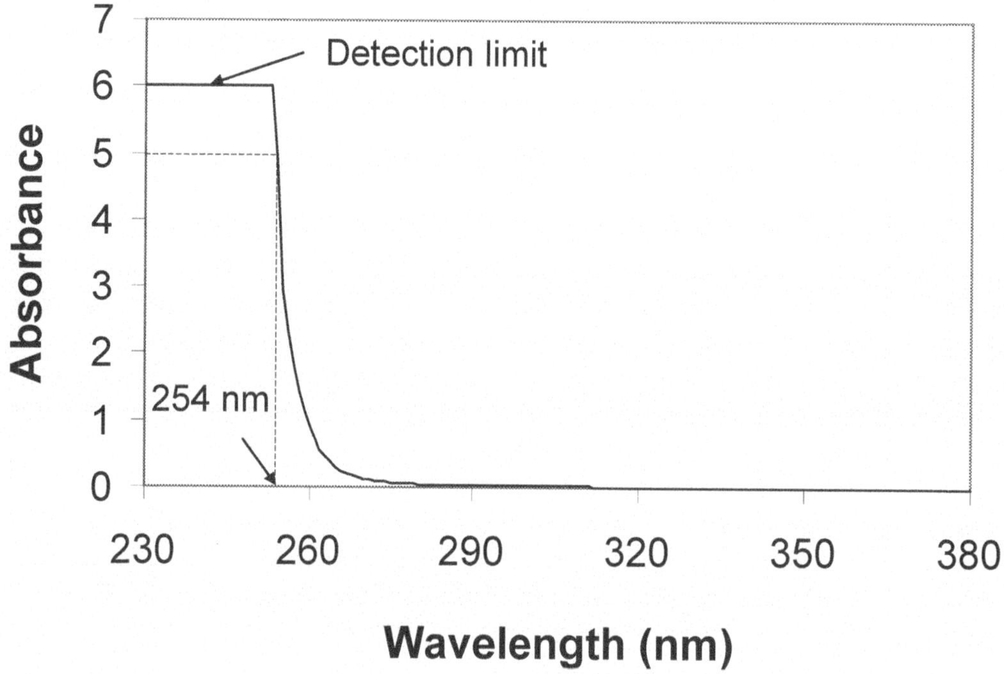

Figure 4-1. Absorbance Spectrum (pathlength = 1.0 cm) of Iodide/Iodate Actinometer After 40-fold Dilution. This acitinometer absorbs UV germicidal radiation effectively, while not absorbing longer wavelength UV and visible radiation.

4.2.2 Reactor Design

4.2.2.1 UV Capillary Flow Reactor

The capillary flow reactor was developed to generate a relatively large volume (on the order of 1 L) of UV-irradiated sample needed for assays described previously (e.g., respiration tests and membrane filtration technique), so that the wastewater samples could receive a controlled, quantifiable UV dose within a workable exposure time.

The capillary flow reactor was designed based largely on the output from a point source summation (PSS) model. The UV intensity at a specific location was estimated using this model. After the capillary flow reactor was constructed, UV dose delivery was further validated by actinometry and biodosimetry.

The reactor design is illustrated schematically in Figure 4-2. The radiation source chosen for this system was an LP mercury arc lamp, with an effective arc length of 42 cm and characteristic output wavelength of 254 nm. The lamp was housed in the center of a wooden box to prevent uncontrolled exposure to radiation. A fan was installed in the box to circulate air and promote rapid thermal equilibration of the lamp. A collimator was mounted at the geometric center on the sidewall and aimed at the lamp axis to facilitate lamp power estimation (Blatchley, 1997). UV intensity was measured at the end of the collimator with a radiometer (International

Light, Newburyport, MA, IL1400). The radiometer employed a detector (International Light, SEL 240) equipped with a filter (T2NS254) to allow only radiation at the wavelength(s) of interest to be received by the photocell.

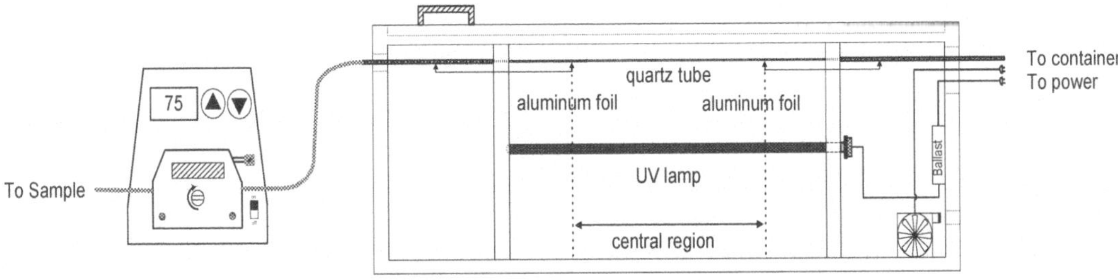

Figure 4-2. Schematic Representation of UV Capillary-flow Reactor.
The liquid to be irradiated was pumped through a capillary quartz tube oriented parallel to a single LP mercury arc lamp.

Sample irradiation was achieved by pumping through a capillary quartz tube (ID=2 mm, OD=4 mm), positioned parallel to the UV lamp. The sample flow rate (exposure time) was controlled using a digital peristaltic pump. All tubes connecting to the capillary quartz tube were covered with aluminum foil. Aluminum foil was also used on both ends of the capillary quartz tube to limit solution irradiation to a region characterized by nearly homogeneous UV intensity. Flow-through actinometry experiments were conducted on the UV reactor under different flow rates, and at different tube positions.

As described previously, the UV intensity distribution of this reactor was calculated using the PSS model (Figure 4-3). All other reactor characteristics required for the PSS calculation were geometric, and were measured using conventional instrumentation for determination of system dimensions. At a fixed radial position, the intensity distribution was nearly homogeneous near the central region of the irradiated zone (Figure 4-4). Based on this finding, the reactor was designed to allow sample irradiation only in the central region. Irradiation outside this region was prevented by covering the relevant segments of the capillary tube with aluminum foil (see Figure 4-2). High and nearly homogeneous intensity of the central (uncovered) region of the capillary-flow UV reactor, together with the liquid pumped through the system, facilitated dose estimation and significantly reduced sample collection time for samples compared with a conventional UV collimated-beam system, in which irradiation can only be accomplished in a small (<10 mL) batch reactor.

4.2.2.2 Point Source Summation Model

The intensity field of the capillary flow UV reactor was calculated by using a PSS approach. For any given receptor site within an irradiated zone, the intensity contributed from hypothetical point sources within the system can be expressed as (Figure 4-3):

$$I_i(R,Z) = \sum_{i=1}^{n} \frac{\frac{P}{n}}{4\pi\rho_i^2} \exp\left[-\left(\sigma_q t_q\right)\frac{\rho_i}{R}\right] \quad (4\text{-}1)$$

where, I: radiation intensity at wavelength λ (mW/cm^2)
R: radial distance from lamp axis to receptor site (cm)
Z: longitudinal distance from one end of the lamp (cm)
P_i: lamp output power at wavelength λ (W)
n: number of point sources
σ_q: absorbance coefficient for quartz jacket (1/cm)
σ_W: absorbance coefficient for water (1/cm)
t_q: thickness of capillary quartz tube (cm)
ρ_i: distance between the ith point source and receptor site

The intensity field along the quartz at a specific radial distance from the UV lamp can be calculated using PSS. Figure 4-4 shows the result; corresponding parameter values used for calculation are listed as follows:

Lamp power=8.8 W
Lamp length=42 cm
Number of point sources=42
Quartz jacket transmittance=0.95
Quartz tube wall thickness=0.1 cm

The intensity decreased at both ends of the quartz tube and remained (nearly) homogeneous in the central region. The maximum variation of intensity within the exposure zone was estimated to be 7.6% and 8.4% at R=10 cm and 14 cm, respectively. The intensity values in the homogenous zone were 3.3 mW/cm^2 and 2.07 mW/cm^2 at R=10 and 14 cm, respectively. These intensity outputs are higher than a typical output of a collimated beam system (~0.1 mW/cm^2 to 0.15 mW/cm^2); thus this system can be used to deliver a nearly uniform UV dose to a relatively large sample (on the order of liters) in a short period of time.

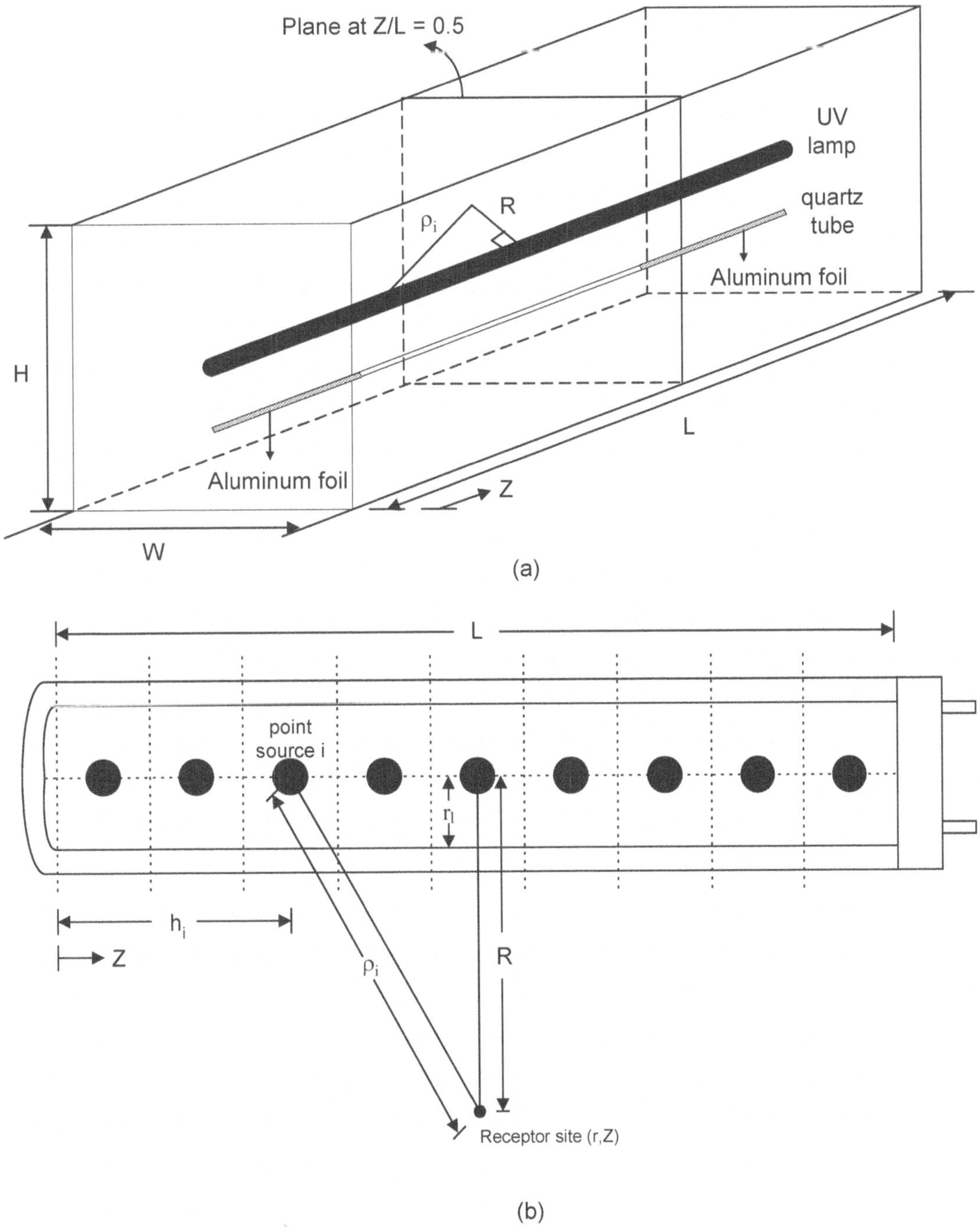

R: radial distance from lamp axis to receptor site (cm)
ρ_i: distance from i^{th} point source to receptor site
r_l: quartz jacket inside radius (cm)
t_q: quartz tube thickness (cm)
h_i: Cartesian coordinate defining horizontal distance for i^{th} point source

Figure 4-3. Schematic Illustration of UV Capillary Flow Reactor (a) and Point Source Concept of UV Lamp (b).

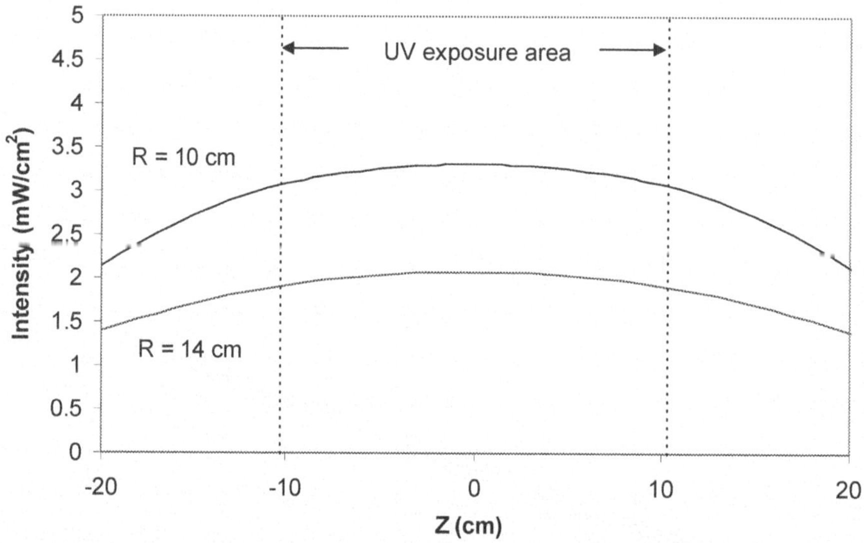

Figure 4-4. UV Intensity Distribution Along Quartz Tube at Different Radial Distance from Lamp Axis, as Calculated by PSS.

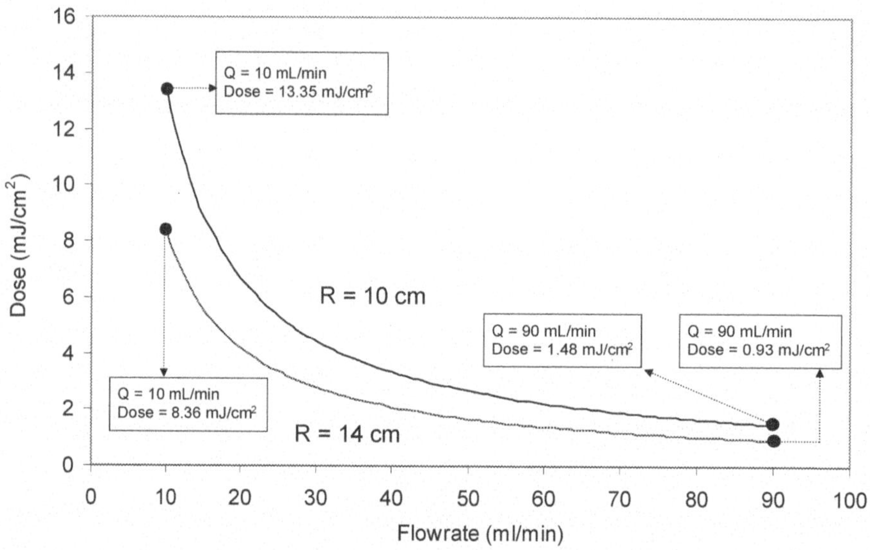

Figure 4-5. UV Dose (by PSS Model) as a function of Flow Rate and Tube Position.

UV dose delivery by the capillary reactor was adjusted by changing the distance between the quartz tube and UV lamp or by changing flow rate of the sample. Figure 4-5 shows a typical operation curve estimated by PSS. For flow rates ranging from 10 mL/min to 90 mL/min, the output doses were estimated to range from 1.48 mJ/cm^2 to 13.36 mJ/cm^2 and 0.93 mJ/cm^2 to 8.36 mJ/cm^2 at R=10 cm and 14 cm, respectively. This system also offers flexibility for higher dose output by connecting more than one tube in series at the same radial distance. Dose output was inversely related to the flow rate. For practical purposes, it is not recommended to use a flow rate <10 mL/min, since the output dose changes greatly with the flow rate under this flow rate region. From this operation curve, it is possible to control the output dose precisely, especially under sufficiently high flow rate. Precise delivery of large UV doses was accomplished by using multiple passes of a water sample through the reactor.

4.2.3 Experimental Procedures

4.2.3.1 Actinometry

Chemical actinometry provides a direct quantifiable chemical measurement of available electromagnetic radiation energy in a photochemical reactor, and is an easy, fast, accurate, and potentially low-cost method for estimating UV dose. It is a radiation-induced chemical process in which UV energy, when absorbed by a photosensitive chemical, will induce a chemical change. Possible chemical changes accomplished in actinometry include photoreduction (e.g., potassium ferrioxalate (Harris et al., 1987), photohydrate reaction (e.g., uridine [Linden and Darby, 1997]) or other measurable change in the parent chemical (Gahr et al., 1995).

An iodide/iodate actinometer was used for quantification of dose delivery in the capillary flow reactor. Photochemical characteristics of the actinometer were determined using a collimated beam system with a quantifiable UV dose. The actinometer was used in the capillary flow reactor at different radial distances between the quartz tube and the UV lamp. Spectrophotometric measurements of iodate production were then compared against the results from a collimated beam system for purposes of dose estimation. The effects of non-uniform intensity and non-uniform velocity within the capillary tube were assumed to be unimportant and were ignored.

4.2.3.2 Biodosimetry

Biodosimetry tests were conducted on the capillary flow reactor using undisinfected wastewater effluent samples collected from Facilities A, B, C, and D. Two types of biodosimetry tests were conducted; the first test used a single UV dose on different wastewater effluents, and the second used several UV doses on a single wastewater effluent. UV exposures for all biodosimetric assays were conducted on a collimated beam system and a capillary flow reactor. Prior to each biodosimetry test, intensity was measured using a radiometer and a collimator that was connected to the capillary flow reactor; the intensity was then used for the lamp output power calculation (an inverse PSS model). The flow rate required to achieve the designated UV dose was calculated by the PSS model.

A collimated beam was used in the biodosimetry test to generate a UV dose-response relationship of microorganisms. Biodosimetry test results from the capillary flow reactor were converted to "equivalent dose" based on observed inactivation. Before each biodosimetry test, samples were equilibrated to room temperature. Biodosimetry tests were carried out using membrane filtration as described in Section 3.2.1.2.

4.3 Capillary Flow UV Reactor

Performance of the capillary flow reactor was validated by actinometry and biodosimetry, and was compared with the results from the collimator. For the collimator and capillary reactor systems, an LP mercury vapor lamp was employed as the radiation source to provide essentially monochromatic irradiation (λ=254 nm).

4.3.1 Actinometry

An iodide/iodate actinometer was used for quantification of UV dose (Rahn, 1997). This actinometer can absorb UV germicidal radiation effectively with little or no absorbance of ambient light. Upon UV exposure, the iodide ion was transformed to the triiodide ion by the following reaction:

$$3I^- \underset{\leftarrow}{\xrightarrow{UV}} I_3^- \text{ (352 nm)} + 2e^- \tag{4-2}$$

As indicated in reaction (4-2), excess electrons will be produced in this reaction, and these will drive the reaction in the reverse direction and make the actinometer solution unstable. To minimize this interference, iodate is added in the actinometer solution to serve as electron scavenger, with reaction stoichiometry defined as follows:

$$IO_3^- + 6e^- + 6H^+ \rightarrow I^- + 3H_2O \tag{4-3}$$

In this study, starch solution was also added to the actinometer solution immediately following UV exposure to further enhance stability. Starch will bind with I_3^- to form an amylose-iodine (AI) complex or amylopectin-iodine (API) complex (Davis et al., 1994). The concentration of the stock starch solution was 500 mg/L and 100 μL of starch solution was added into each 8 mL sample of actinometer solution. The addition of starch did not change actinometer absorbance in the wavelength range 320 nm to 380 nm. Applied UV dose was determined by quantification of I_3^- formation, which can be quantified by using its absorbance value at 352 nm. A typical result of iodide/iodate actinometer before and after UV exposure (with addition of starch solution) is shown in Figure 4-6.

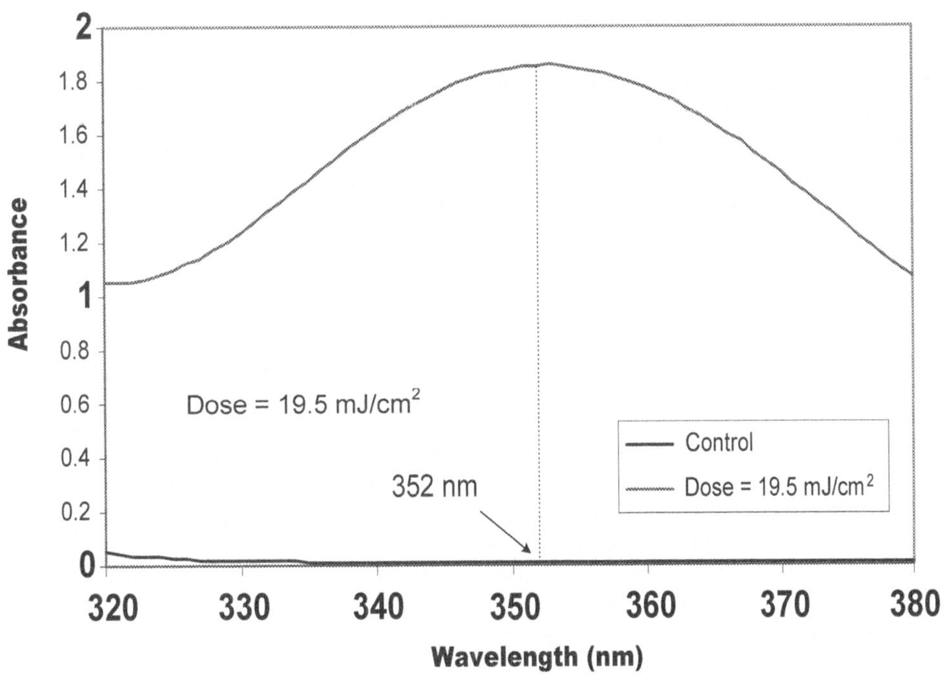

Figure 4-6. Absorbance Spectrum of Iodide/Iodate Actinometer Before and After UV_{254} Irradiation. Control sample represents the same sample without UV irradiation.

The stability of this actinometer was tested by addition of starch solution after UV exposure. An initial UV dose of 10 mJ/cm^2 was imposed on the actinometer solution using a collimated beam and a well-mixed shallow batch reactor. Solution absorbance at 352 nm was recorded with time following UV exposure. The result showed that absorbance decreased from 0.9487 to 0.9465 (~0.2%) within the first five minutes. This suggests that the photoproduct was acceptably stable under these conditions. Furthermore, the actinometer underwent oxidation to form triiodide ion under low pH conditions, as illustrated below:

$$6I^- + O_2 + 4H^+ \rightarrow 2I_3^- + 2H_2O \qquad (4\text{-}4)$$

To limit interference caused by proton-enhanced oxidation, the actinometer was kept at pH 9.25 using a borate/phosphate buffer.

4.3.2 UV Collimator With Shallow Dish

A well-mixed shallow dish reactor was used with a UV collimator to characterize the chemical properties (quantum yield and extinction coefficient) of the iodide/iodate actinometer. The radiation source was an LP mercury vapor lamp, which emits nearly monochromatic radiation at a characteristic wavelength of 254 nm. The quantum yield is defined as

$$\varphi_{254} = \frac{\text{number of moles produced}}{\text{number of photons absorbed}} \qquad (4\text{-}5)$$

The photochemical reaction rate, ν, can be described as

$$\upsilon_{254} = I_{a,254}\varphi_{254} \qquad (4\text{-}6)$$

Where $I_{a,254}$ is the rate of photon absorbance per unit volume at 254 nm.

According to the Beer-Lambert law,

$$I_{a,254} = I'_{0,254}(1 - e^{-2.3\varepsilon_{254}Sl}) \qquad (4\text{-}7)$$

For the iodide/iodate actinometer, the measured absorbance at 254 nm was 4.94 after 40 fold dilution using the same pH buffer. This implies a very large value of $\varepsilon_{254}S$ for this actinometer. Therefore, the solution may be considered to be opaque at this wavelength, and equation (4-6) can be simplified to

$$I_{a,254} = I'_{0,254} \qquad (4\text{-}8)$$

In this experiment, $I'_{0,254}$ was measured using a radiometer under the collimator system. Because of the strong absorbance of this actinometer at 254 nm, the iodide/iodate actinometer absorbs essentially all the incident photons; the flux of incident photons can be derived from the radiometer reading. From equation (4-4):

$$\varphi_{254} = \frac{\text{number of moles produced}}{\text{number of photons absorbed}} = \frac{C_1 V_1}{\text{number of photons absorbed}} \quad (4\text{-}9)$$

According to Beer's Law

$$ABS_1 = \varepsilon_{352} C_1 l \quad (4\text{-}10)$$

Therefore, equation (4-8) can be rewritten as

$$\varphi_{254} = \frac{ABS_1 \times V_1}{\varepsilon_{352} \times \text{number of photons absorbed} \times l} \quad (4\text{-}11)$$

and

$$\varphi_{254} \times \varepsilon_{352} = \frac{ABS_1}{\text{number of photons absorbed} \times l / V_1} \propto \frac{ABS_1}{E_{abs} / V_1} \quad (4\text{-}12)$$

In the collimator system with a shallow well-mixed reactor, E_{abs} can be determined by radiometry since the area of the dish reactor is measurable. The product of φ_{254} and ε_{352} can thus be determined from a collimator system by using the slope of the dose-response curve. Typical results of the dose response of this actinometer under this experiment set are illustrated in Figure 4-7. The repeatable results showed a linear relationship between UV dose and absorbance at 352 nm. This linearity also implies that the iodide/iodate actinometer follows a zero-order reaction under the conditions used in these experiments, and the slope of the dose-response relationship is linearly proportional to the product of φ_{254} and ε_{352}. This actinometer was sensitive to temperature, so the information regarding the product of φ_{254} and ε_{352} was required for normalization among different batch experiments.

4.3.3 UV Collimator with a Flow-through Quartz Tube

Lamp output power (254 nm) was estimated in each experiment using a measurement of collimated UV intensity at a well-defined position, relative to the lamp. A collimator of length (H) 25 cm and opening diameter (d) 4.3 cm was installed on the sidewall of the reactor. A radiometer detector was positioned at one end of the collimator, with its surface perpendicular to the axis of collimation and centered within the collimator opening. Lamp output power was calculated using the PSS model:

$$I_{reading} = \sum_{z=z_i}^{z_f} \frac{\frac{P}{n}}{4\pi L(\Delta^2 + z^2)} \qquad (4\text{-}13)$$

Where $\Delta = S + t + H$
See Figure 4-9 for a description of terms.

Actinometry tests were conducted on the capillary flow reactor under different flow rates (Q=11.5 – 91.0 mL/min) and different radial distances between the quartz tube and lamp (R=10 cm and 14 cm). Under steady-state conditions, actinometer conversion (which is directly related to delivered dose) was expected to be directly proportional to irradiation time, or inversely related to flow rate. The resulting absorbance of photochemical product from the iodide/iodate actinometer showed excellent linearity and repeatability (Figure 4-10). In the case of R=10 cm, a slight yield of absorbance was observed. A UV collimator was used with flow-through quartz tube to provide a homogeneous, stable, and measurable UV intensity. The purpose of this experiment was to obtain a standard curve for comparison with the dose-response relationship observed under the extreme low flow condition.

Numerical predictions and results from the actinometry tests were compared. Absorbance of the photochemical products from the iodide/iodate actinometer in the capillary flow reactor was translated to absorbed energy by using the following equations:

$$C_2 = \frac{ABS_2}{\varepsilon_{352}\, l} = \frac{\text{number of moles produced}}{V_2} \qquad (4\text{-}14)$$

From equation (4-4), equation (4-13) can be rewritten as:

$$C_2 = \frac{ABS_2}{\varepsilon_{352}\, l} = \frac{\varphi_{254} \times \text{number of photons absorbed}}{V_2} \qquad (4\text{-}15)$$

So,

$$\text{number of photons absorbed} = \frac{ABS_2 \times V_2}{\varphi_{254} \times \varepsilon_{352} \times l} \qquad (4\text{-}16)$$

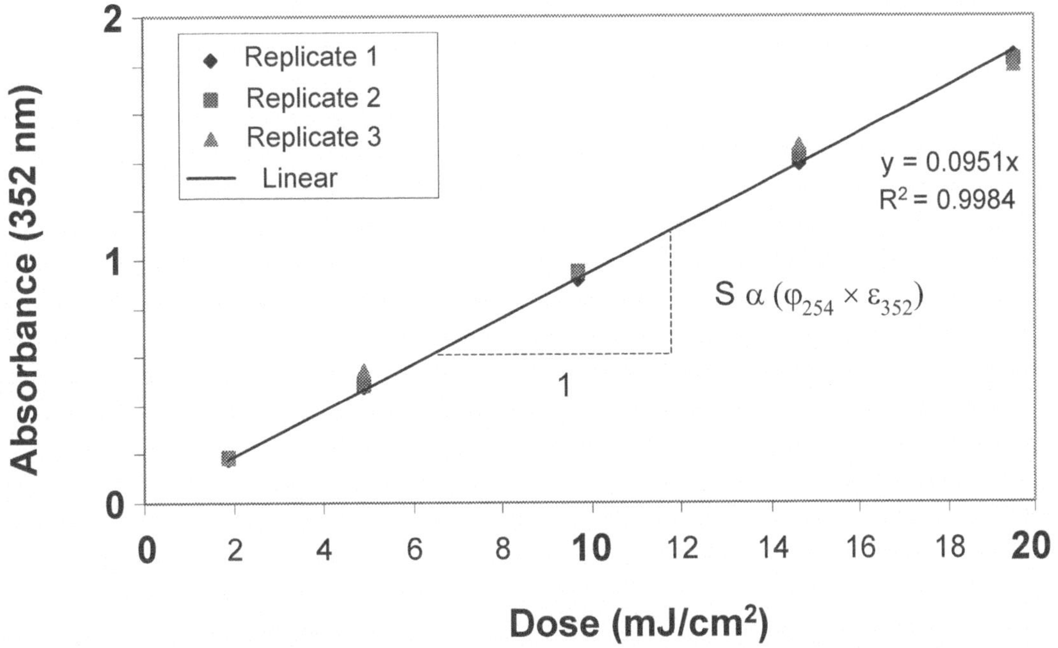

Figure 4-7. Dose-response Behavior of Iodide/Iodate Actinometer Under a UV Collimated Beam in a Well-mixed, Shallow Dish Reactor.

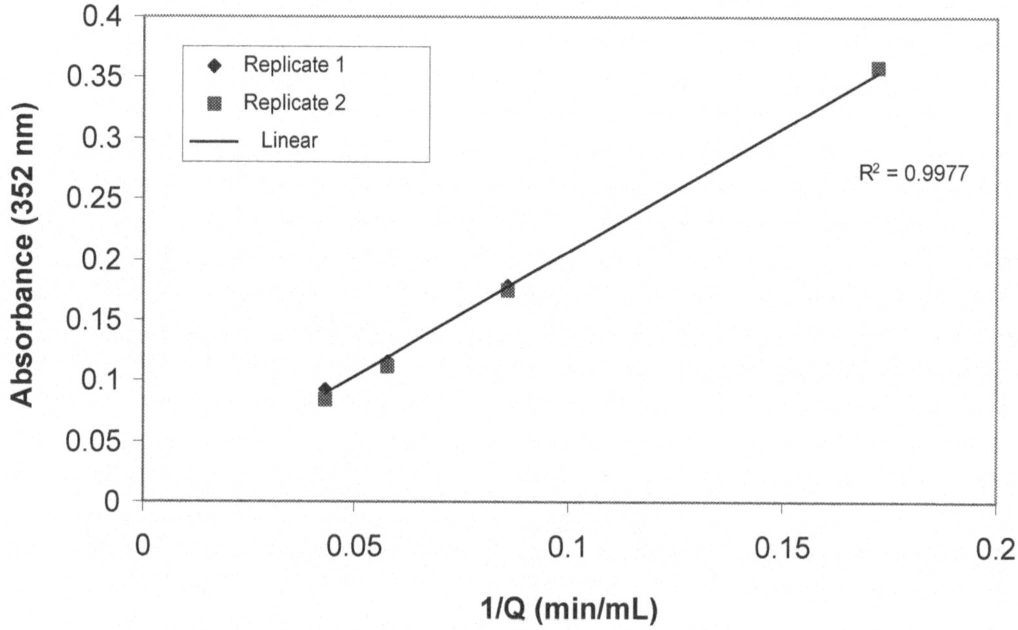

Figure 4-8. Dose-response Behavior of Iodide/Iodate Actinometer Under a UV Collimated Beam in a Flow-through Quartz Tube.

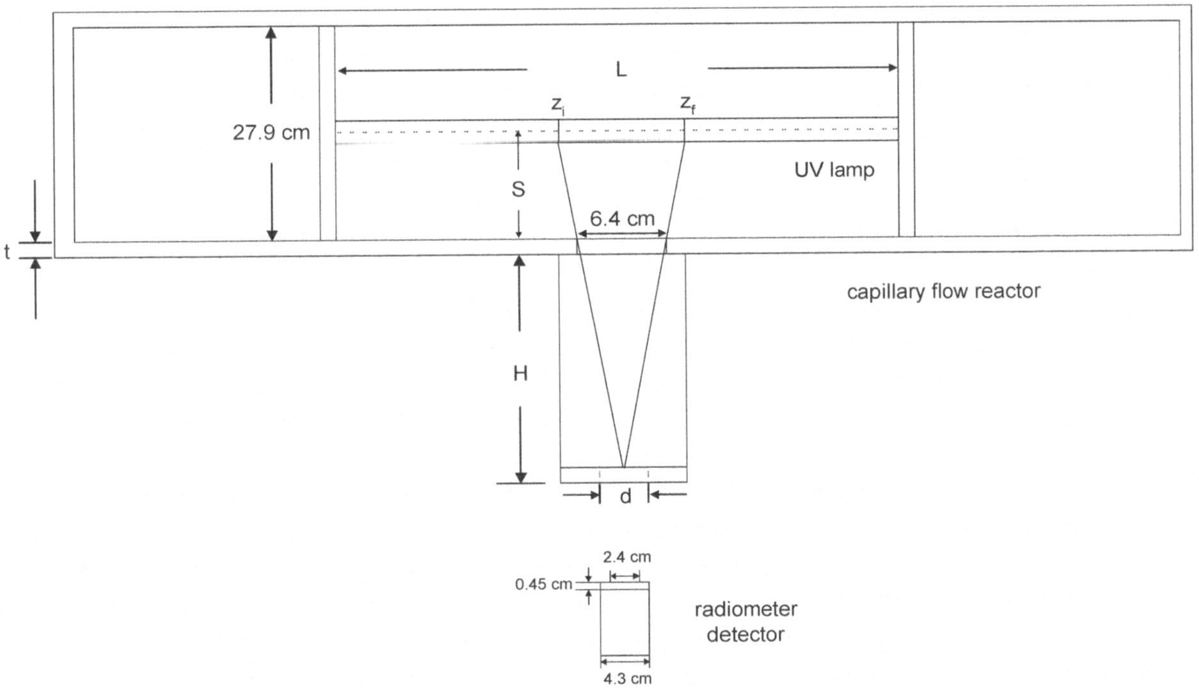

Figure 4-9. Apparatus Used to Estimate Lamp Output Power.
The collimater and detector were placed in the geographic center for measurement of intensity. Measurements of UV intensity were conducted by placing the surface of the detector flush with the end of the collimator.

The product of φ_{254} and ε_{352} was estimated from collimated beam system experiments (described above). The exposure volume (V_2) in the capillary flow reactor was defined by the product of the capillary cross-sectional area and irradiated length. From the observed absorbance of the photochemical product, the number of photons absorbed was calculated using equation (4-15).

The incident dose was estimated using the PSS model. The equivalent incident dose was estimated by comparison with the dose-response behavior of the flow-through quartz tube under the collimated beam system. Figure 4-11 shows the relation between the calculated dose and equivalent incident dose. Because of the strong absorbance of the iodide/iodate actinometer, virtually all incident photon energy was absorbed by the actinometer. Under these circumstances, the "available dose" was equivalent to the "absorbed dose." The slope decreased when the quartz tube got closer to the lamp (smaller R) (Figure 4-11). This may be attributable to increasing importance of reflection loss from the longitudinal direction that was not taken into account in the PSS model.

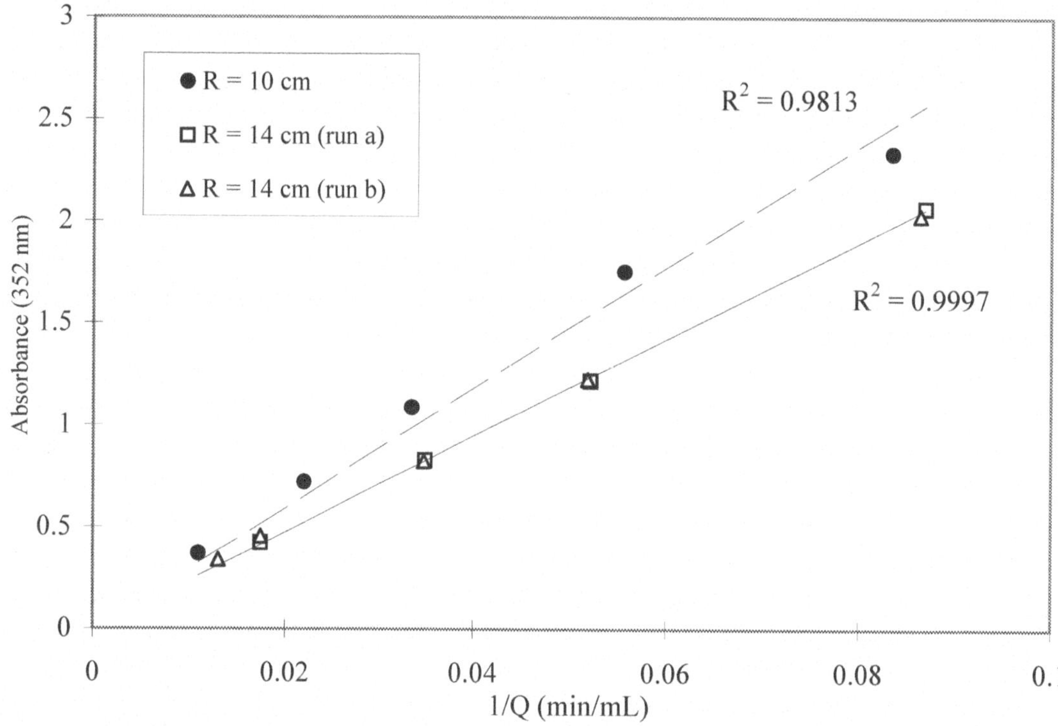

Figure 4-10. Absorbance Measurements in Iodide/Iodate Actinometer Solution, Collected from Capillary Flow Reactor Under Steady-State Conditions, as a Function of (Inverse) Flow Rate.

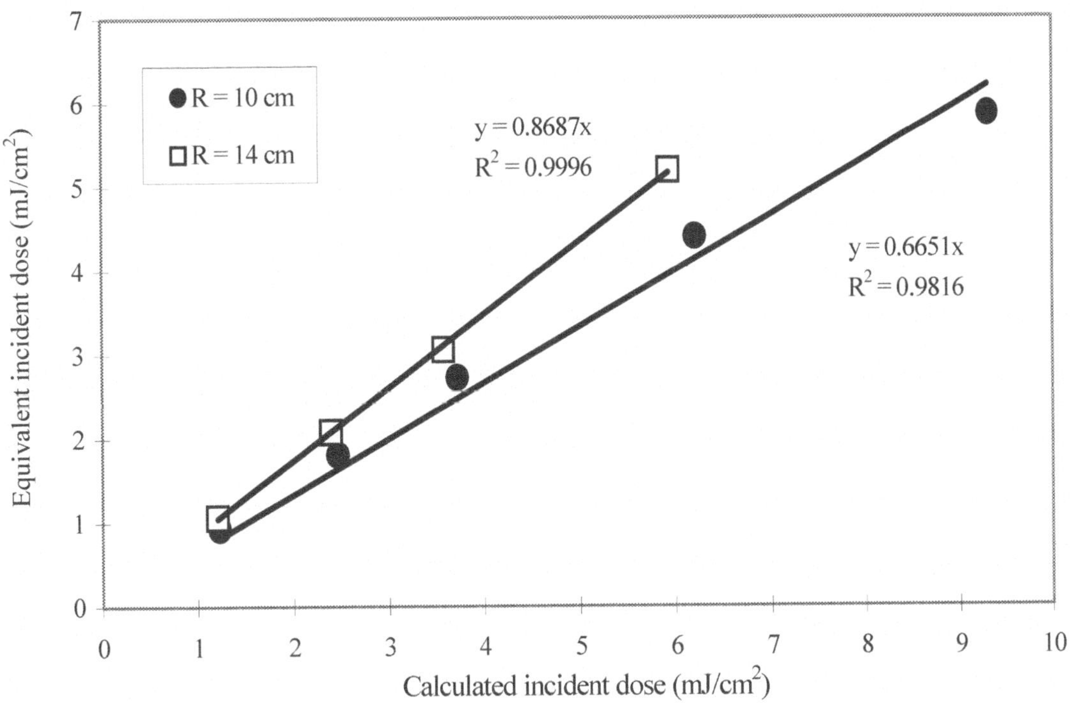

Figure 4-11. Comparison of Equivalent Dose (By Actinometry) and Calculated Dose (By PSS Model).

4.3.4 Biodosimetry

Biodosimetry was also used for evaluation of the performance of the capillary flow reactor. Some of the characteristic advantages and disadvantages of photochemical reactor characterization by actinometry assay and biodosimetry assay are listed in Table 4-1. In Wastewater effluent samples collected from Facilities A, B, C, and D were subjected to UV irradiation by using the collimator and capillary flow reactor in order to compare equivalency. After UV exposure, each sample was subjected to membrane filtration and incubation for quantification of FC concentration. For collimator experiments, different UV doses were applied to the same wastewater effluent sample, and the UV intensity was measured by radiometry. The average intensity was calculated by

$$I = I_0 e^{-\alpha x} \qquad (4\text{-}17)$$

$$I_{avg} = \frac{1}{H}\int_0^H I_0 e^{-\alpha x}\,dx \qquad (4\text{-}18)$$

$$\rightarrow \quad I_{avg} = \frac{1-e^{-\alpha H}}{\alpha H} I_0 \qquad (4\text{-}19)$$

Where,
I_0 = incident intensity as measured by radiometry
H = sample depth
α = transmittance coefficient.

Table 4-1. Advantages and Disadvantages of UV Reactor Characterization by Chemical Actinometry and Biodosimetry.

	Advantage	Disadvantage
Actinometry	- Highly reliable, high R^2 - Repeatable - Less interference - Fast - Easy - Routine measurements are possible	- Usually works at high dose range - Some actinometers are not very stable - Relatively expensive - Need to prepare actinometer solution
Biodosimetry	- Sensitivity depends on initial concentration and species of microorganism, sensitivity can be high - Good linearity at low dose - Inexpensive	- Usually low linearity at high dose due to deviations from first-order behavior in dose-response curve - More interference - Less repeatable - Relatively low reliability, low R^2 - Slow - Complicated (membrane filtration assays, counting…) - Routine measurements are not practical

For the capillary reactor experiment, the operation condition (flow rate) required to achieve the target UV dose was calculated using the PSS model. In this experiment, a narrow range of target UV doses (11.82 mJ/cm^2 to 12.88 mJ/cm^2) was used for all wastewater effluent samples. The reason for choosing this dose range was that the dose-response curve for the target organism tends to become less linear due to tailing at higher UV doses. The resulting $\log_{10}$ inactivation value obtained from the capillary flow reactor was then translated to an equivalent UV dose by comparing with the dose-response curve obtained from the collimator experiment using regression analysis. However, when tailing occurred, only the nearest region was used for regression rather than using the entire UV dose range in order to reduce variation.

Figures 4-12 and 4-13 illustrate the results of biodosimetry. The calculated dose by PSS can reasonably predict the actual disinfection efficiency of the collimator system. The variation between the calculated dose and predicted dose may be attributable to the slight tailing behavior in the dose-response curve of some samples. From Figure 4-13, the transmittance of wastewater effluent sample ranged from 15.4% to 82.4%. The ratio P' in Figure 4-13 is defined as

$$P' = \frac{\text{Equivalent dose of biodosimetry test}}{\text{Calculated dose by PSS model}} \qquad (4\text{-}20)$$

The calculated dose by the PSS model and the equivalent dose by biodosimetry were comparable within this transmittance range.

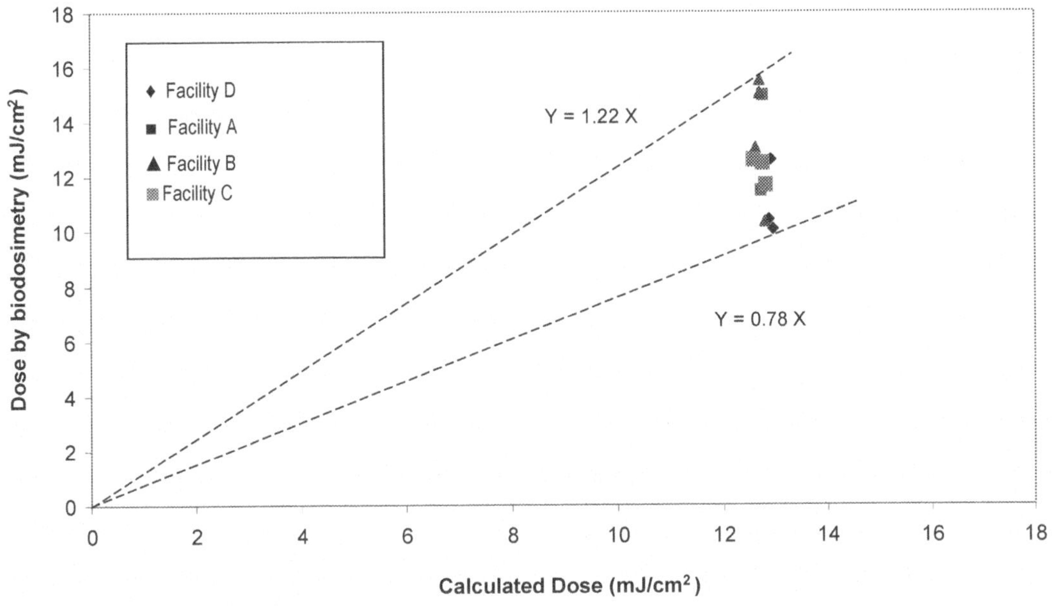

Figure 4-12. Results of Biodosimetry Test Using Capillary Flow Reactor.
The calculated UV dose was estimated by PSS model using the measured intensity from radiometer. The dose of biodosimetry was obtained by comparing the result with dose-response curve from a collimated-beam system.

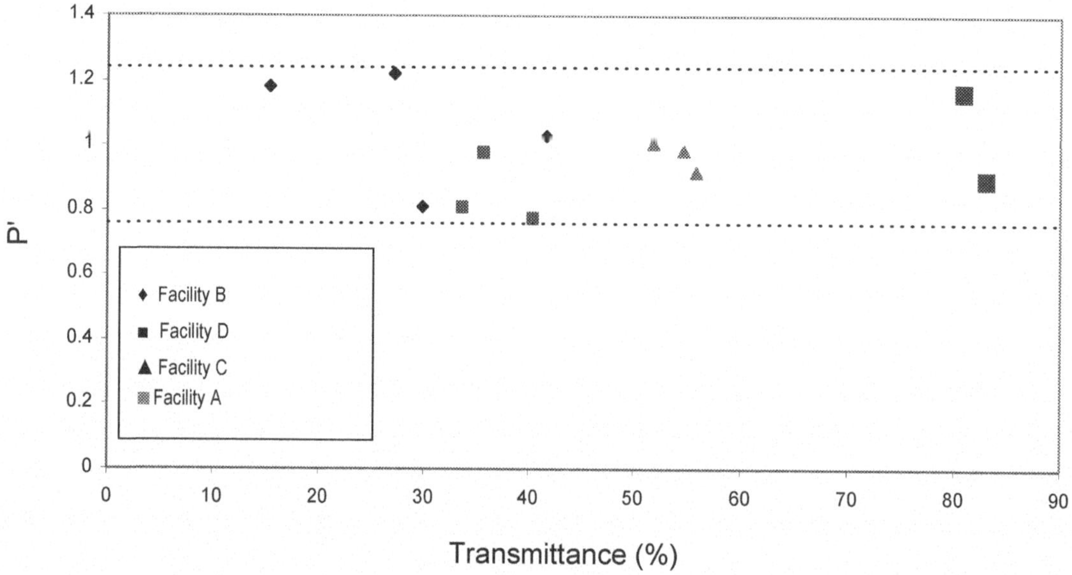

Figure 4-13. Results of Biodosimetry Test Using Capillary Flow Reactor.
Four types of wastewater effluents with different transmittance were used in this test. P' is defined as the ratio of equivalent dose of biodosimetry test to calculated dose by PSS model.

4.4 Summary and Conclusions

The results of experiments and numerical analyses of the capillary flow UV reactor system demonstrate that it is capable of UV irradiating water in a continuous-flow mode. With the flow rates that are possible with this system, relevant doses of UV radiation can be delivered to sample volumes on the order of liters within a practical time frame—typically tens of minutes. It is possible to treat similar aqueous volumes using a collimated-beam system; however, this would require substantially longer periods to accumulate such a volume (several hours to days). Over such a period, substantial changes in water quality may occur. Likewise, relatively large volumes of UV-irradiated water can be collected from conventional continuous-flow reactors. However, accurate and reliable characterization of UV dose delivery in these reactors is highly suspect based on available methods.

CHAPTER 5.0

Culture- and Non-culture-based Methods for Assessment of Bacterial Responses to Chlorination and UV Irradiation

5.1 Introduction

Recent progress in the development of in situ, intracellular fluorescent probes has facilitated the rapid assessment of bacterial physiology in biofilms and planktonic populations, without cultivation. Efforts to describe the physiological activities of bacterial populations have been hindered by the inability of culture-based techniques to recover a significant proportion of bacteria present in a sample (Amann et al., 1995) and by traditional dependence on methods that yield community averages of physiological indices, such as oxygen consumption, and radiolabeled substrate incorporation into macromolecules. The advantages of using fluorescent intracellular probes include the direct assessment of 1) physiological status; 2) specific metabolic activities; 3) gene expression; and 4) total cell densities (McFeters et al., 1995). The ability to provide expeditious results pertaining to the presence and physiological activity of specific bacteria has direct application in a range of systems, including the monitoring of drinking water, wastewater, untreated and treated process waters (e.g., pharmaceuticals), and foods. Intracellular fluorescent probes have been used for assessing physiological activity after disinfection of biofilm-associated bacteria (Huang et al., 1995; Stewart et al., 1994; Yu and McFeters, 1994) and planktonic cells (Lisle et al., 1998; 1999).

5.2 Objectives

Experiments were conducted in which the responses of wastewater effluent bacteria to chlorination or UV irradiation were characterized using culture- and non-culture-based methods. Specific objectives included:

- Quantification of the abundance of FCs, enterococci, and total culturable bacteria in treated wastewater samples.
- Characterization of the efficacy of chlorine and UV radiation for disinfection of treated wastewater samples by calculating the log_{10} reductions in FCs, enterococci, and total culturable bacteria.
- Statistical comparison of disinfection efficacies of chlorination and UV irradiation in terms of bacterial responses to these disinfection processes.
- Assessment of the applicability of culture- and non-culture-based methods in the assessment of disinfection efficacy.

5.3 Materials and Methods

5.3.1 Sample Collection and Processing

Pairs of 1.0 L effluent samples were collected during a 15-month period from selected wastewater treatment facilities, as listed in Table 5-1. Following collection, the samples were shipped to the laboratory via overnight carrier. Upon receipt, both samples were poured into a sterile 3.0 L flask containing a magnetic stir bar and gently mixed until the sample had equilibrated to 21-23°C. All experiments were conducted at 21-23°C. Once the sample temperature had stabilized, bench-scale (batch) disinfection was conducted, followed by a battery of microbiological assays, as listed in Table 5-1 and described in the following sections.

5.3.2 Disinfection Procedures

Chlorination Free chlorine (HOCl + OCl^-) working solutions were prepared on the day of use using a stock sodium hypochlorite solution (#SS290, Fisher Scientific) and adjusted to a final concentration that allowed 1.0 mL of working solution to be added to the sample to provide a total chlorine residual of 2.0 mg/L or 3.0 mg/L as Cl_2 (Table 5-1). Free chlorine was added to 1.0 L of sample that had been aseptically transferred to a sterile 2-L beaker that contained a magnetic stirring bar. The sample was mixed at a slow but constant rate for 40 minutes. Free and total chlorine residuals were determined using a Hach DR800 colorimeter (Hach Co., Loveland, CO) and a preset program for total and free chlorine residual analysis using a modification of the DPD method, as described by the manufacturer. At each sample collection time, a subsample of chlorine-treated sample was removed and transferred to a sterile beaker containing sodium thiosulfate (final concentration: 2% volume/volume [v/v]) to neutralize the residual chlorine.

Ultraviolet Irradiation The UV exposure experiments employed a UV a dose of 10 mJ/cm^2 or 20 mJ/cm^2 (Table 5-1). A flat plate collimated-beam reactor was used to deliver the appropriate UV dose in all experiments (Blatchley, 1997). The reactor used an LP mercury lamp that emitted essentially monochromatic radiation at a characteristic wavelength of 253.7 nm. The collimator was calibrated at Purdue University and delivered an incident UV intensity of approximately 140 $\mu W/cm^2$. A sterile 50 mm diameter glass petri dish, containing a micro-magnetic stirring bar, was used as the reaction vessel. The sample depth in the petri dish was limited to 1 cm, which allowed 18 mL of sample to be irradiated per dose exposure. Each sample was continuously stirred during the exposure period. The total volume of sample required for each UV exposure experiment ranged from 250 mL to 500 mL. To accumulate these final volumes, multiple 18 mL volumes were exposed at the appropriate UV dose and then transferred to a sterile beaker containing a magnetic stirring bar. The beaker was wrapped and covered in foil to minimize exposure to ambient light and photoreactivation. This collective volume was continuously stirred at a slow but constant rate. The number of sample volumes exposed ranged from 15 to 30. Samples were subjected to UV irradiation under the collimated beam for three or six minutes to achieve a dose of 10 mJ/cm^2 or 20 mJ/cm^2, respectively.

5.3.3 Culture-based Microbiological Assays

Fecal coliform recovery Samples from all wastewater treatment facilities were assayed for the presence of FCs (Table 5-1) using mFC agar and the membrane filtration method (American Public Health Association [APHA], 1998). Oxyrase was also used to treat a parallel set of membrane filters for each FC sample. Oxyrase is a liquid enzyme system that selectively reduces the concentration of reactive oxygen species that have been shown to be detrimental to bacteria (Puig et al., 1998).

Oxyrase was included to determine if the reactive oxygen species generated during the disinfection processes enhanced the injury to exposed bacteria and inhibited recovery. Following filtration, the membrane filter was transferred onto a sterile absorbent pad saturated with Oxyrase for 15 seconds. The filter was then transferred to mFC agar. All mFC agar plates were sealed in a plastic bag and incubated in a 44.5°C water bath for 22-24 hours. Colonies that developed a blue coloration were counted as FCs.

Table 5-1. Summary of Sample Locations and Completed Microbiological Assays.

Facility Identifier	Sample Collection Date	Chlorine Dose (mg/L as Cl_2)	Time (min)	UV Dose (mJ/cm^2)	FC	Ent	TCB	TDC	Bacterial Membrane Potential		Bacterial Injury Assay (%)
A	12/6/2000	2.0	40	10	x		x	x	x	x	
	6/26/2001	2.0	40	20	x		x	x	x	x	x
B	4/17/2001	2.0	40	20	x		x	x	x	x	x
	6/7/2001	2.0	40	20	x		x	x	x	x	x
	9/19/2001	2.0	40	20	x		x	x	x	x	x
	10/4/2001	2.0	40	20	x		x	x	x	x	x
E	3/6/2002	3.0	40	20	x		x	x	x	x	x
	3/20/2002	3.0	40	20	x		x	x	x	x	
D	3/10/2001	2.0	40	10	x	x	x	x	x	x	x
	4/3/2001	2.0	40	20	x	x	x	x	x	x	x
	8/23/2001	2.0	40	20	x	x	x	x	x	x	x
	9/6/2001	2.0	40	20	x	x	x	x	x	x	x
C	8/8/2001	2.0	40	20	x	x	x	x	x	x	x
	10/17/2001	2.0	40	20	x	x	x	x			x
	11/7/2001	2.0	40	20	x	x	x	x			x

Enterococci recovery As described in Section 3.4, Facilities C and D discharge their treated effluents to a marine environment. Samples from Facilities C and D were assayed for the presence of enterococci (Table 5-1) using the two-step membrane filtration method that utilizes the selective media mE and EIA agars (APHA, 1998). Oxyrase was also used to treat a parallel set of membrane filters for each enterococci sample. Following filtration, the membrane filter was transferred onto a sterile absorbent pad saturated with Oxyrase for 15 sec. The filter was then transferred to the first medium in the two-step method, mE agar. All mE agar plates were incubated at 41°C for 18 to 24 hours, prior to transferring the membrane filters to EIA agar for 20 minutes to confirm the presence of enterococci. Colonies that developed a black precipitate in the media were counted as enterococci.

Total culturable bacteria TCB counts were obtained by plating samples onto R2A agar (Reasoner and Geldreich, 1985), with and without Oxyrase supplementation. Oxyrase was added according to the manufacturer's recommendations. Plates were inoculated using a spread plate technique (APHA, 1998) and incubated at 21°C to 23°C for 14 days, with counts recorded on days 2, 5, and 14. Duplicate plates were inoculated for each sample.

Injury assay Storage, shipping, and disinfectant-induced injury to cell membranes were assessed using a differential plate count assay (Pyle et al.,1995; Terzieva and McFeters, 1991; Zaske et al., 1980). This assay uses tryptone lactose yeast extract (TLY) and TLYD (TLY with 0.1% deoxycholate) agars. The membrane disruptive action of the detergent deoxycholate is enhanced in cells with compromised membranes. Duplicate plates were inoculated as described for the TCB assay and inoculated using a modified drop plate technique (Hoben and Somasegaran, 1982).

5.3.4 Non-Culture Based Microbiological Assays

Total direct cell counts For total direct cell counts (TDC), 10.0 mL samples were removed from the sample flask and added to a sterile 15 mL tube that contained 150 µL of formalin (37% v/v). The sample was gently mixed and stored at 4°C until it was counted. DAPI (4',6-diamidino-2-phenylindole), a nucleic acid stain, was dissolved in reagent grade water and filter sterilized. DAPI was added to each sample at a final concentration of 10 µg/mL, incubated at 4°C overnight and counted the next day (Porter and Feig, 1980). Samples were filtered through 0.2 µm pore size, 25-mm diameter black polycarbonate membrane filters (GTPB, Millipore Corp., Billerica, Massachusetts). Filters were transferred to glass slides and mounted by placing a coverslip on which 20 µL of a sterile 50:50 solution of glycerol and phosphate buffered saline (PBS; pH 7.5) had been placed, on top of the filter. All TDC counts were conducted using a Nikon Eclipse 600 epifluorescent microscope equipped a 100-W mercury light source, 100x oil immersion objective (UV-F) and an EF4UV filter cube. Duplicate filters were prepared from each sample and a minimum of 20 fields, as determined by an ocular reticule and at least 400 cells were counted per filter (Fry, 1990, 1993; Vives-Rego, J. et al., 2000).

Physiological activity (membrane potential) assay All samples were assayed for the presence of a membrane potential prior and following disinfection (Table 5-1). The presence of a membrane potential in bacterial cells prior to and following disinfectant exposure was determined using Rhodamine 123 (Rh123) (Lisle et al., 1998). Rh123 is a lipophilic cationic stain that is preferentially accumulated in cells that possess a membrane potential (Haugland, 1996). Rh123 was prepared in TE (10 mM TRIS, 1 mM EDTA, pH 8.0) and used at a final concentration of 13.0 µM. Cell suspensions were subjected to membrane filtration as previously described. With the membrane filter still in the filtration funnel, 600 µL of Rh123 was added as an overlay and allowed to incubate in the dark at 21°C to 23°C for 10 minutes.

The absence of a membrane potential in the same bacterial suspension was determined using bis-(1,3-dibutylbarbituric acid) trimethine oxanol ($DiBAC_4$) (Lisle et al., 1998). $DiBAC_4$ (APHA, 1998) is a lipophilic anionic stain that is excluded from cells possessing a membrane potential (Haugland, 1996). $DiBAC_4$ was prepared in reagent grade water and used at a final concentration of 20.0 µM. Cell suspensions were filtered and stained as described for Rh123. Rh123 and $DiBAC_4$ filters were counted using the microscope, B-2E filter cube, and criteria described previously.

5.3.5 Statistical Analyses

All microscopic and plate count data were corrected for applicable dilution and microscope correction factors. A value of 1 was added to each value [i.e., (x + 1)]; all values were then $\log_{10}$ transformed and confirmed as collectively following a normal distribution prior to any statistical manipulation. Pearson moment correlation coefficient, one-way ANOVA and t-test analyses (assuming equal variances) were performed on the transformed data. The graphed TDC, FC and enterococci data were converted to geometric means and the asymmetrical 95% confidence intervals (CI) calculated for the respective means. All statistical analyses were performed at $\alpha=0.05$ (Fry, 1993; Sokal and Rohlf, 1981). Statistical analyses were performed using Minitab, release 13.32 (Minitab, Inc., State College, PA).

5.4 Results and Discussion

5.4.1 Effects of Disinfection on TDC

TDCs are used to determine the total biomass or bioload associated with a sample. As such, TDC measurements provide a gross measure of microbiological burden in a water sample.

TDC values at each sample site may change between sample dates due to changes in microbiological quality of the WWTP influent. These types of changes occur on time scales that range from minutes to days, and can be due to changes in flow rates, storm water contributions, and industrial and domestic activities. However, TDC values should have remained relatively constant within each sample during the disinfection assays used in this study, as the doses of chlorine and UV were not great enough to promote a significant reduction in cell numbers due to cell lysis. Figures 5-1 supports this assumption, as there were no statistically significant ($p>0.10$) differences between the TDC values for untreated, chlorinated, and UV irradiated samples on each of the sample dates for the respective WWTPs. Additionally, the pooled data from all TDC untreated, chlorinated, and UV irradiated samples were significantly correlated ($R^2=0.983$, $p<0.001$), and there were no significant differences ($p= 0.986$) between the data sets. Collectively, these data indicate that the disinfectants used in this study at respective doses and contact times do not promote the removal of bacterial cells through lysis regardless of the geographic location, seasonal effects (i.e., temperature, sunlight), or treatment processes.

However, when comparing all the TDC values for samples collected from each of the facilities, statistically significant ($p < 0.001$) differences were observed among the TDC measurements based on sample dates. These results also illustrate an important issue that should be considered when assessing disinfection efficacy in wastewater treatment systems: the measurement of bioload is dynamic and difficult to accurately predict.

Though TDC counts provide information regarding total bioload, current regulations require the simple monitoring of an indicator bacterial group, such as FCs, to characterize microbiological quality of wastewater treatment facility effluent and disinfection efficacy. The indicator bacterial groups (e.g., FCs, enterococci) are present in the total bacterial population in relatively small percentages. Table 5-2 summarizes the percentage of the TDC that FCs comprised for each sample collected. These percentages range from 0.002% (Facility D) to 29.2% (Facility B). In general, the FC fraction of the TDC for Facilities A, C, and E were relatively consistent and low. In contrast, samples from Facility B were characterized by a relatively high FC fraction and greater variability. Collectively, the data in Table 5-2 indicate that FCs were a minor and variable component of the total bioload and that adjusting disinfectant doses and exposure times for control of bacteria based on $\log_{10}$ reductions of this indicator group should be done with caution.

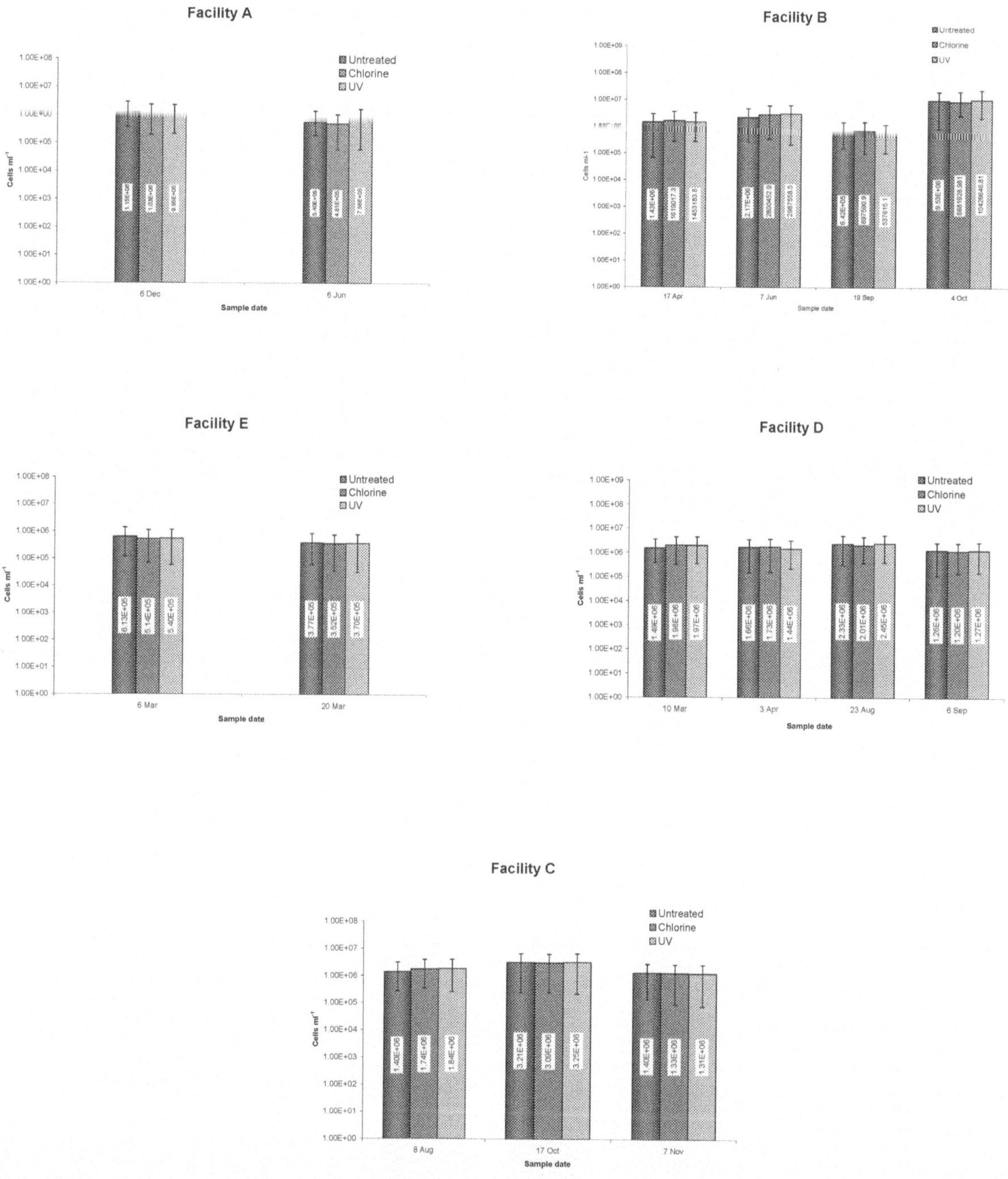

Figure 5-1. TDC Abundances for Effluent Samples from Facilities A-E.

Table 5-2. Percentage of TDCs Prior to Disinfection Treatment.[1]

Facility Identifier	Sample Collection Date	Fecal Coliforms	Enterococci	TCB[2]
A	12/6/2000	0.053		50.4
	6/26/2001	0.103		95.2
B	4/17/2001	1.41		24.5
	6/7/2001	29.2		74.2
	9/19/2001	16.7		68.6
	10/4/2001	0.163		37.2
E	3/6/2002	0.236		37.4
	3/20/2002	0.702		4.70
D	3/10/2001	0.002	2.915	87.4
	4/3/2001	18.3	0.787	107.3
	8/23/2001	1.52	0.834	5.15
	9/6/2001	13.6	0.135	62.9
C	8/8/2001	0.045	0.022	86.8
	10/17/2001	0.244	0.003	68.1
	11/7/2001	0.271	0.092	73.7

[1] All values are expressed as percentages.

[2] Values calculated after 14 days of incubation

All samples analyzed in this study had substantial concentrations of suspended particulates that were easily seen during the TDC microscopic analyses. These particulates were not crystalline solids (e.g., sand) but instead were rather large, amorphous and had a "fuzzy" appearance with an open matrix. DAPI-stained bacterial cells were always seen within these particles. These particulates influence the culture-based data in two ways. First, regardless of how many bacteria are embedded within or on a particulate only a single colony (i.e., colony forming unit [CFU]) will form on a media plate. This will cause an underestimation of the respective count values. Second, bacteria associated with particulates can be substantially more resistant to chlorine and UV disinfection than planktonic, individual cells because of physical protection from the disinfectants.

5.4.2 The Effects of Disinfection on Fecal Coliform

Fecal Coliform Abundance State and federal regulations require monitoring of the indicator group of bacteria, FC, in wastewater treatment facility effluent. These regulations are designed to assess the microbiological quality of the effluent and to indirectly provide an indication of the potential for microbial contamination following contact or ingestion of the effluent or receiving waters. Additionally, the efficacies of disinfectants used in wastewater treatment processes are commonly evaluated using the FC indicator group. Figure 5-2 depicts the FC abundances among sampling dates. There were no statistically significant differences ($p>0.08$) in the FC abundances between sampling dates at Facility A, Facility E, and Facility C . Facilities B and D had statistically significant ($p<0.05$) differences in FC abundances for samples collected on June 7, 2001, and March 10, 2001, respectively. These dates also coincide with a respective increase and decrease in the percentage of FC per TDC at these two facilities (Table 5-2). Also, there was no significant correlation between TDC and FC values (Table 5-3) when using pooled data from all samples, and these data were significantly different ($p<0.001$). These data indicate that while the TDC counts remain relatively consistent between sample dates, the percentage that each type or group of bacteria (e.g., FC) contributes to the TDC fluctuates. A description of the factors that may contribute to this type of variability has been presented previously.

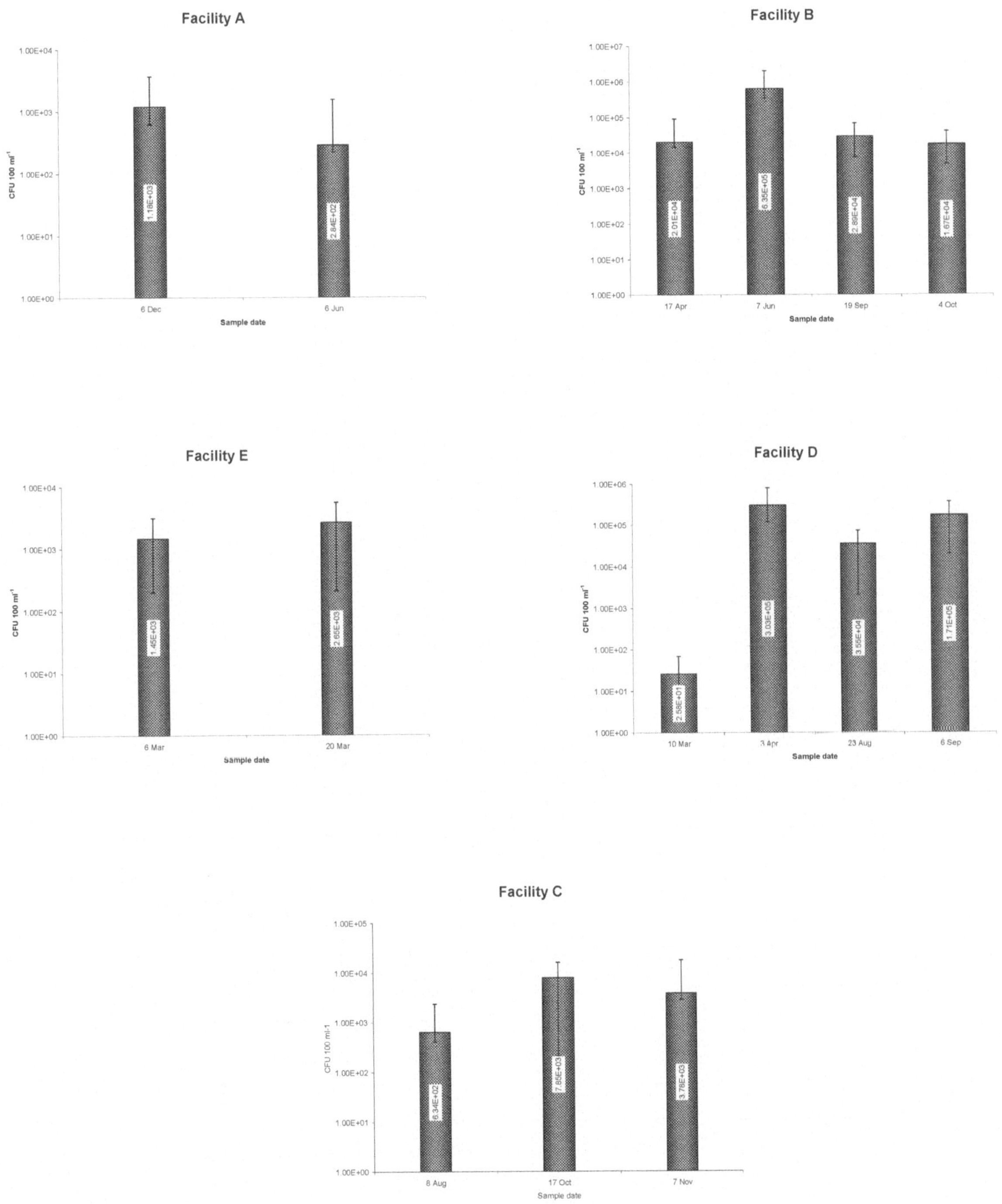

Figure 5-2. FC Abundances for Effluent Samples from Facilities A-E.

Table 5-3. Percentage of TCB Counts Prior to Disinfection Treatment.[1,2]

Facility Identifier	Sample Collection Date	Fecal Coliforms	Enterococci
A	12/6/2000	0.11	
	6/26/2001	0.10	
B	4/17/2001	5.75	
	6/7/2001	39.4	
	9/19/2001	6.58	
	10/4/2001	0.44	
E	3/6/2002	0.63	
	3/20/2002	15.0	
D	3/10/2001	0.002	3.40
	4/3/2001	17.4	0.73
	8/23/2001	29.6	16.2
	9/6/2001	21.6	0.21
C	8/8/2001	0.05	0.02
	10/17/2001	0.36	0.01
	11/7/2001	0.37	0.12

[1] All values are expressed as percentages.

[2] Values calculated after 14 days of incubation.

Reduction in Culturable Fecal Coliform Following Disinfection FC removal or reduction, calculated as the difference between the $\log_{10}$ values of viable FC concentration prior to and following treatment is a commonly used parameter for characterization of disinfection efficacy. The reduction of culturable microorganisms in wastewater by disinfectants, such as chlorine and UV radiation, is influenced by several factors (Bitton, 1994). Chlorine-based disinfection of wastewater can be influenced by disinfectant concentration, contact time, pH, temperature, and physiological status of the target microbes (e.g., laboratory grown, starved, stressed). Turbidity, color, inorganic and organic nitrogenous compounds, iron, manganese, hydrogen sulfide, and total organic carbon have been shown to consistently and negatively influence chlorine disinfection efficiency. UV disinfection efficiency is similarly influenced by turbidity, color, physiological status of the target microbes, and photoreactivation.

When comparing the FC $\log_{10}$ reduction values following disinfection with chlorine, there was some variability between samples from each facility (Figure 5-3). For each of the facilities except Facility C, bacterial inactivation among the samples showed substantial variability. There appears to be no seasonal explanation for this variability; rather it is likely that changes in the microbiological, chemical, and physical components of the wastewater streams were responsible for the observed variations in disinfection efficacy.

The UV disinfection $\log_{10}$ reduction values for each facility showed trends similar to those of chlorine (Figure 5-3). Using two sample t-tests to compare chlorine and UV $\log_{10}$ reduction values for Facility B ($p=0.837$), Facility D ($p=0.988$), and Facility C ($p=1.000$), no statistically significant differences were found in the $\log_{10}$ reduction values between the two disinfectants (Note: there were too few data points for samples from Facilities A and E to perform a test of data normality). These results indicate that for the conditions under which these experiments were conducted, there were no significant differences between the efficacy of chlorine and UV radiation as a disinfectant for the reduction of FC; the batch exposure scenarios used for chlorination and UV irradiation were designed to allow compliance with discharge regulations for bacterial quality at each facility. The consistency in measured bacterial inactivation responses is important for comparison of the behavior of the two disinfection methods.

5.4.3 The Effects of Disinfection on Enterococci

Enterococci Abundance The enterococci (Ent) group of bacteria is gaining acceptance as an alternative indicator to FC for the assessment of the microbiological quality of natural waters, especially marine waters. The Ent group is a subgroup of the fecal streptococci that includes *Streptococcus faecalis*, *Streptococcus faecium*, *Streptococcus gallinarum*, and *Streptococcus avium* (APHA, 1998; Godfree et al., 1997). The preference for using Ent over FCs in marine waters is because Ent have been shown to be relatively resistant to coastal marine water salinity levels (Davies et al., 1995; Godfree et al., 1997; Pinto et al., 1999). The Ent were included in this study for samples collected from Facilities C and D because both discharge into estuarine and coastal marine waters. The motivation for these measurements was to determine if Ent were present in the wastewater streams at a concentration that would allow them to be used as an indicator of disinfection efficacy, and if their abundances closely modeled FC. Figure 5-4 shows the abundances of Ent for each sampling date at the respective sample locations. As with FC, variations in the pooled Ent abundance data did not correlate with TDC and differences between the respective abundances were statistically significant ($p<0.001$). Also, the Ent were present in the wastewater streams at percentages of TDC that were similar to those for FC (Table 5-2). However, the changes in Ent abundances for Facilities C and D (Figure 5-4) did not significantly correlate with changes in the respective FC abundances. There were no statistical differences between these same pooled data for Ent and FC when comparing untreated ($p=0.400$), chlorinated ($p=0.166$), and UV irradiated ($p=0.060$) values. These data suggest that the Ent group are equally prevalent and are recovered at a similar efficiency using culture-based methods as FC. This is an important aspect of the wastewater ecology of this indicator group if it is to be used as an alternative to FC for monitoring of wastewater effluent discharges into marine waters.

Reduction in Culturable Ent Following Disinfection Log_{10} reductions in Ent culturability were used to assess the efficacy of chlorine and UV disinfection. When comparing the $\log_{10}$ reduction values from each sample date at Facility D (Figure 5-5), there was no significant difference ($p=0.564$) between chlorine and UV radiation for inactivation of Ent. Similar ranges of $\log_{10}$ reduction values were obtained for samples from facility C(Figure 5-5), and there was no significant difference ($p=0.419$) between the efficacies of the disinfectants, as with the Facility D samples.

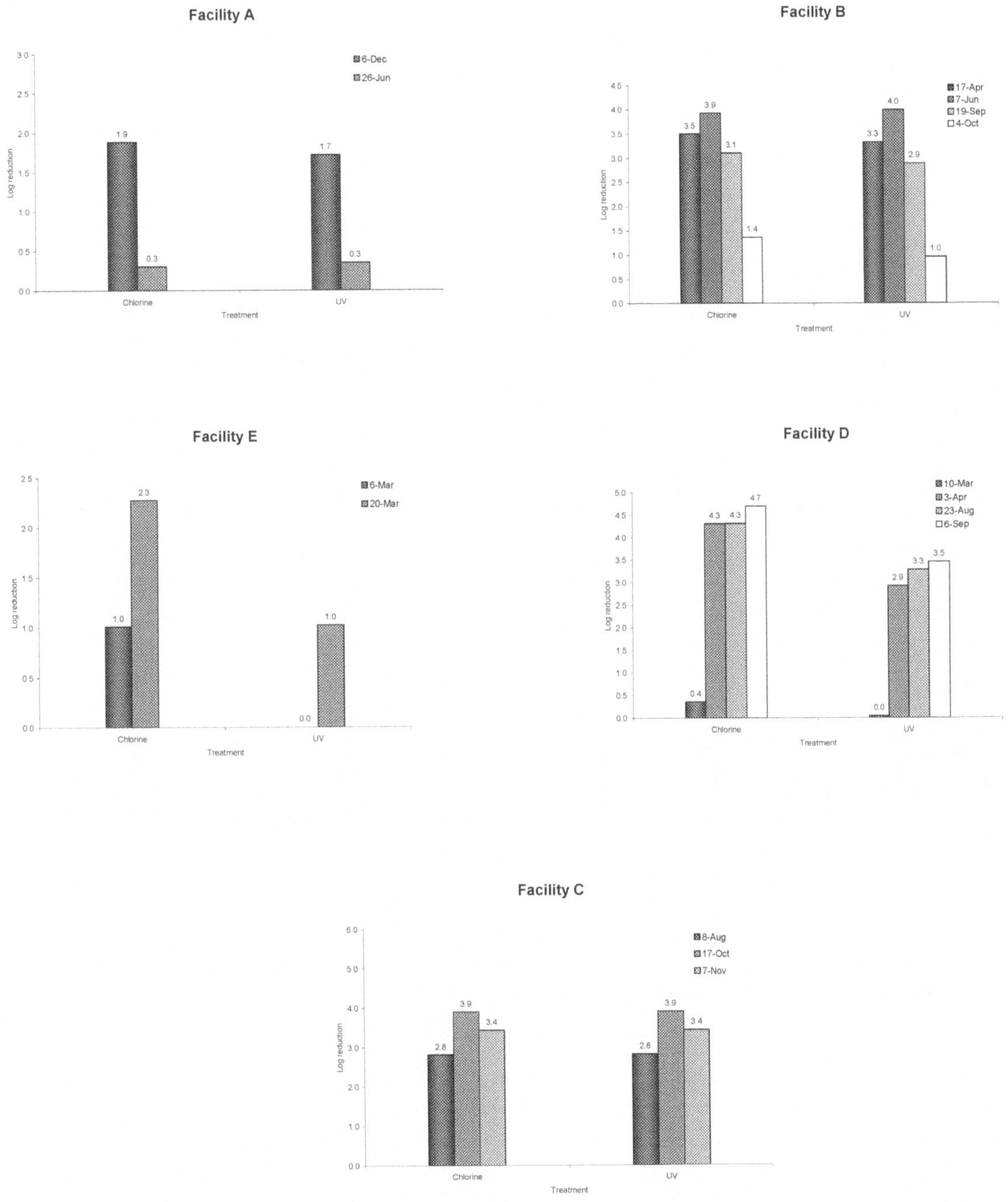

Figure 5-3. Log_{10} Reductions in FC Abundances Following Disinfection for Effluent Samples from Facilities A-E.

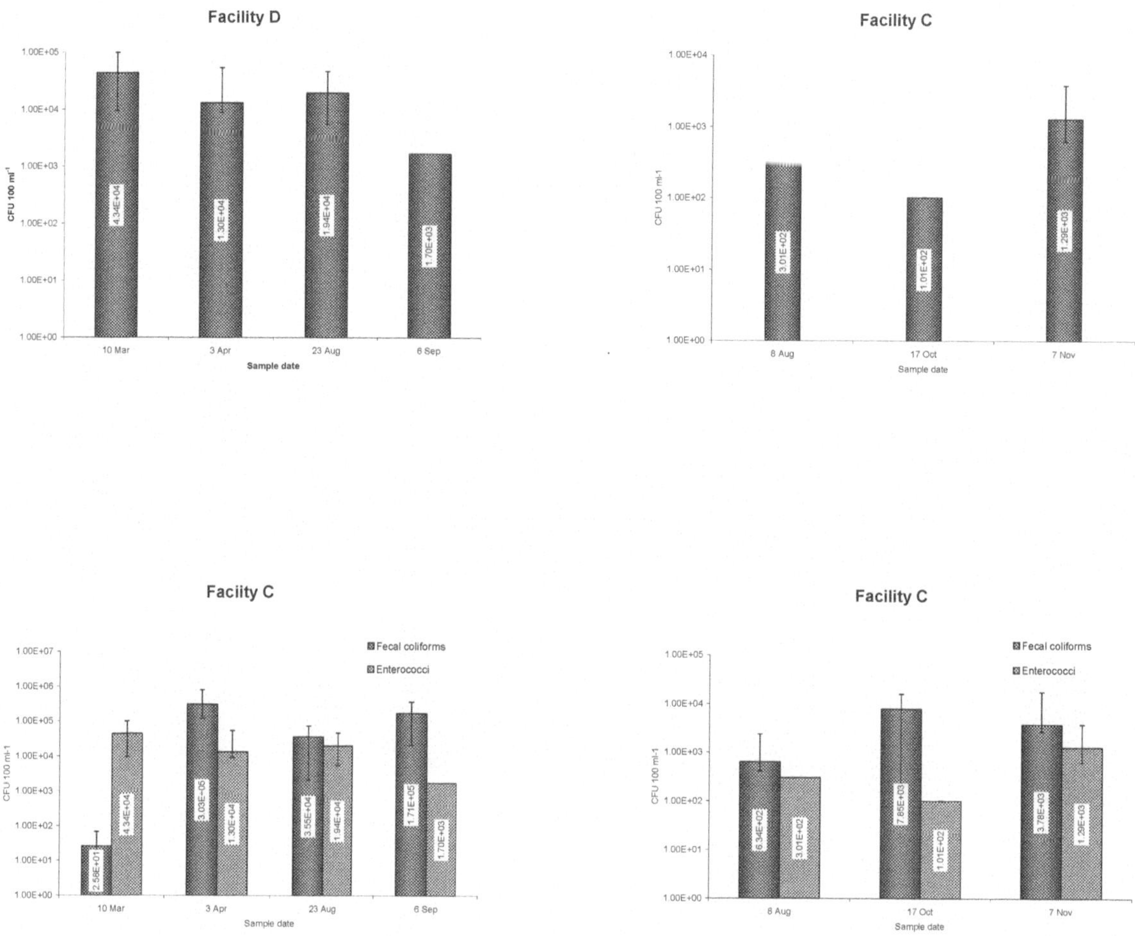

Figure 5-4. Ent Abundances for Effluent Samples Collected from Facilities D and C (top panels); FC and Ent Comparisons for Effluent Samples from Facilities D and C (bottom panels).

When comparing the $\log_{10}$ reduction values for Ent (Figure 5-5) to FC (Figure 5-3) at Facility D (Figure 5-5), there was no significant difference ($p= 0.713$) between the efficacy of chlorine and UV for the inactivation of Ent and FC. However, this same overall equality in the efficacy of both disinfectants was not seen at Facility C (Figures 5-3 and 5-5, collectively shown in Figure 5-5). The disinfection efficacies were statistically equivalent ($p\geq 0.134$) when comparing the $\log_{10}$ reductions values between Ent exposed to UV and FC exposed to chlorine and UV. There were significant differences ($p= 0.029$) between the $\log_{10}$ reduction values when comparing Ent exposed to chlorine to FC exposed to chlorine and

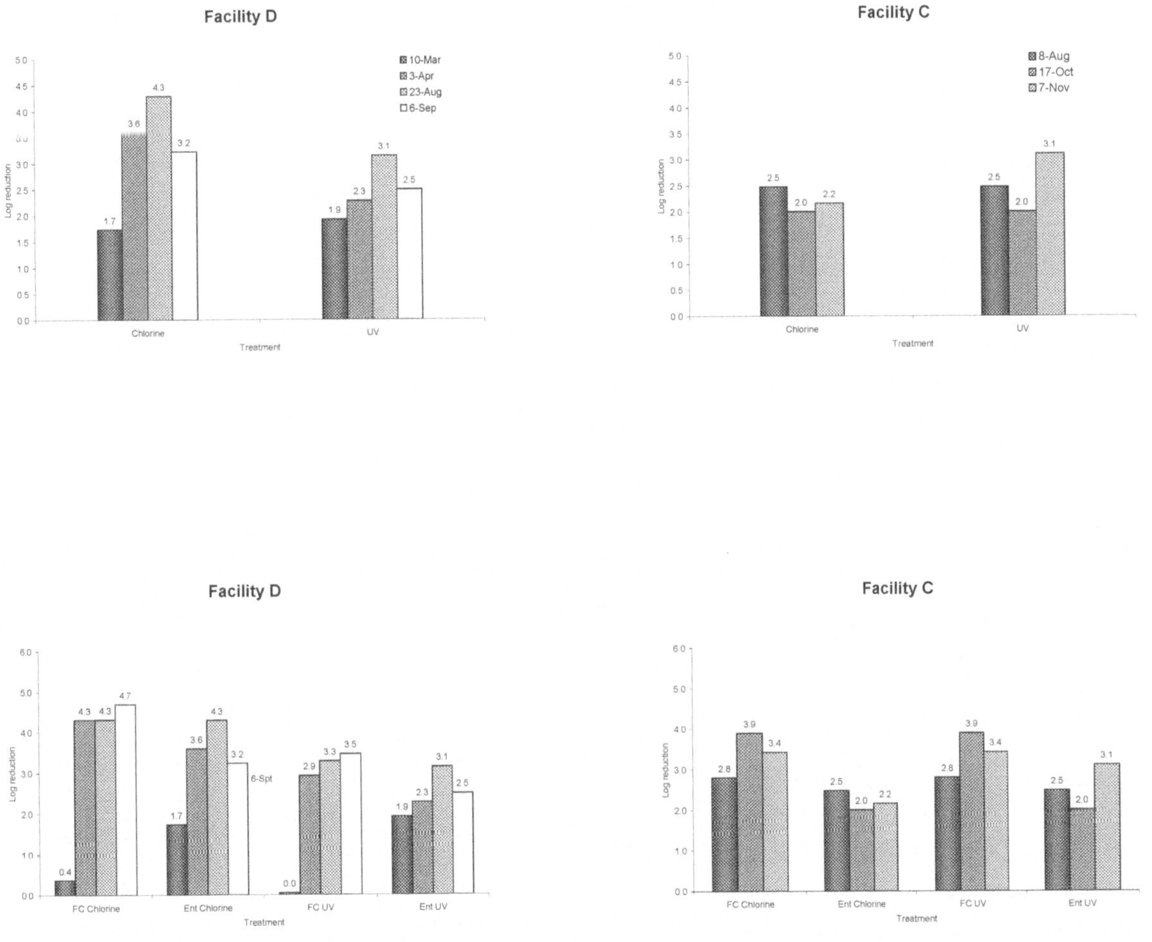

Figure 5-5. Log_{10} Reductions in Enterococci (Ent) Abundances Following Disinfection for Effluent Samples from Facilities D and C (top panels); Log_{10} Reduction Comparisons for FC and Ent in Effluent Samples from Facilities D and C (bottom panels).

UV, with the Ent values being less (i.e., more resistant) than the FC values. These data suggest that the chemistry of the wastewater stream at Facility C may be influencing the chlorination chemistry differently than at Facility D. For example, ammonia interacts with HOCl to form chloramines, which have a significantly reduced oxidation potential relative to HOCl (Berman et al., 1988; Jacangelo et al., 1987; Wolfe et al., 1985). As compared with free chlorine, chloramines often persist in solution for relatively long periods of time; however they require longer contact times than free chlorine to inactivate an equivalent number of microbes (Severin and Churn, 1987). The efficacy of UV irradiation is not directly influenced by ammonia concentrations.

The general trends in these data suggest that Ent were as abundant as FC and these abundances fluctuated within the effluents in a manner similar to FC. Ent and FC bacteria appear to be inactivated by UV radiation at similar rates. However, there was a significant reduction in the efficacy of chlorine in the inactivation of Ent, when compared to FC, at Facility C.

5.4.4 The Effects of Disinfection on Total Culturable Bacteria

Total Culturable Bacteria Abundance TCB are bacteria that can grow on laboratory media at a specific temperature during a given period. The assay used in this study is commonly referred to as the heterotrophic plate count (HPC) (APHA, 1998). The TCB assay was included in this study to provide an indication of the bioload that is culturable and to assess if this method may be applicable for assessment of disinfection efficacies in wastewater treatment facilities. This method does not recover strict anaerobes. The TCB made up from 4.70% (Facility E) to 107.3% (Facility D) of the TDC in the samples analyzed during this study (Table 5-2). When comparing the culture-based methods (i.e., FC, Ent, TCB) FC comprised between 0.05% (Facility C) and 39.4% (Facility B) of the TCB (Table 5-3), while Ent comprised between 0.01% (Facility C) and 16.2% (Facility D) of the TCB (Table 5-3). These data demonstrate again that the indicator groups of bacteria comprised only a small percentage of the total bacterial population present in the wastewater samples. Even the most general and efficient method of culturing the total bacterial populations from these samples performed consistently poorly. It is an established concept in microbial ecology that culture-based methods will not recover all of the bacteria from any type of sample (Amann et al., 1995; Brock, 1987).

The abundances of TCB for the respective sample dates at each facility are illustrated in Figures 5-6. There were no statistically significant differences ($p>0.10$) in the TCB abundances on different sample dates within each facility.

Reductions in TCB Following Disinfection Pooled data from all facilities were used to assess the efficacy of each disinfectant. There were no significant differences ($p=0.052$) between TCB abundances following disinfection with chlorine and UV radiation, but both treatments were significantly different ($p< 0.001$) from TCB abundances in the untreated samples. The TCB abundances in untreated, chlorinated, and UV irradiated samples did not correlate with the FC or Ent abundances in these same samples from the respective facilities. Additionally, there were significant differences ($p<0.001$) between the TCB and the FC and Ent abundances in untreated, chlorinated, and UV irradiated samples. Collectively, these data indicate that the occurrence of and changes in the abundances of TCB is a dynamic process, just like the FC and Ent. However, the TCB populations in the samples assayed were not affected to the same extent by the actions of the disinfectants, as were the FC and Ent. This difference is most likely due significantly greater numbers of TCB than FC and Ent in each sample and because the indigenous bacteria are inherently more resistant, based on culturability, to chlorine and UV disinfection than FC and Ent (Belkin et al., 1999; Matin and Harakeh, 1990; Olson and Stewart, 1987; Russell et al., 1997). Moreover, the TCB assay responds to a broader spectrum of bacteria than the FC or Ent assays. Consequently, it is reasonable to expect that the TCB population will display a broad range of susceptibility to environmental stresses. The TCB assay does not differentiate among bacteria based on their sensitivity to disinfectants. Therefore, the observed behavior of the population of bacteria that respond positively to the TCB assay will be heavily influenced by bacteria that display natural resistance to a form of external stress, such as a disinfectant.

Current regulations that reference the use of this method (i.e., U.S. EPA's Surface Water Treatment Rule, 40 *CFR* 141.74) recommend using a pour plate method and incubating at 35°C for two days. Additional guidelines for the use of this method suggest that increasing the incubation period to five to seven days and lowering the incubation temperature to between 20°C and 28°C will provide conditions for obtaining the highest counts (APHA, 1998). Previous work by Lisle et al. (1998, 1999) has shown that bacterial growth rates on culture media are significantly reduced following exposure to disinfectants and that prolonged incubation at room temperatures significantly increased recovery efficiencies. It is worth noting that incubation periods for FC and Ent cannot be extended, as both media used in these assays would be overgrown with non-FC and non-Ent bacteria within five days. Additionally, the respective methods have been standardized for regulatory applications, and alterations to these methods technically invalidate any resulting data.

In this study, the incubation period at 21-23°C was extended to 14 days, with counts conducted on days 2, 5, and 14. Figures 5-7 (a-b), 5-8 (a-d), 5-9 (a-b), 5-10 (a-d), and 5-11 (a-c) illustrate TCB recovery for each of the samples collected from each facility. There are common trends among the data sets. The first is the data for the untreated samples was consistently greater than the data for the chlorinated and UV irradiated samples throughout the incubation period. Second, there was a general increase in TCB values during the incubation period. Another common feature was the relatively large range that the 95% CIs for the chlorine and UV treated samples included, with generally greater interval ranges occurring in the data sets for the day 2 counts. This was due to the negative influence that sub-lethal stress from exposure to chlorine and UV radiation had on overall metabolism and subsequent growth (Camper and McFeters, 1979; Huang et al., 1995; Lisle et al., 1998, 1999; McFeters, 1990, 1997; McFeters et al., 1986, 1993, 1999; McFeters and Camper, 1983; Terzieva and McFeters, 1991, 1992).

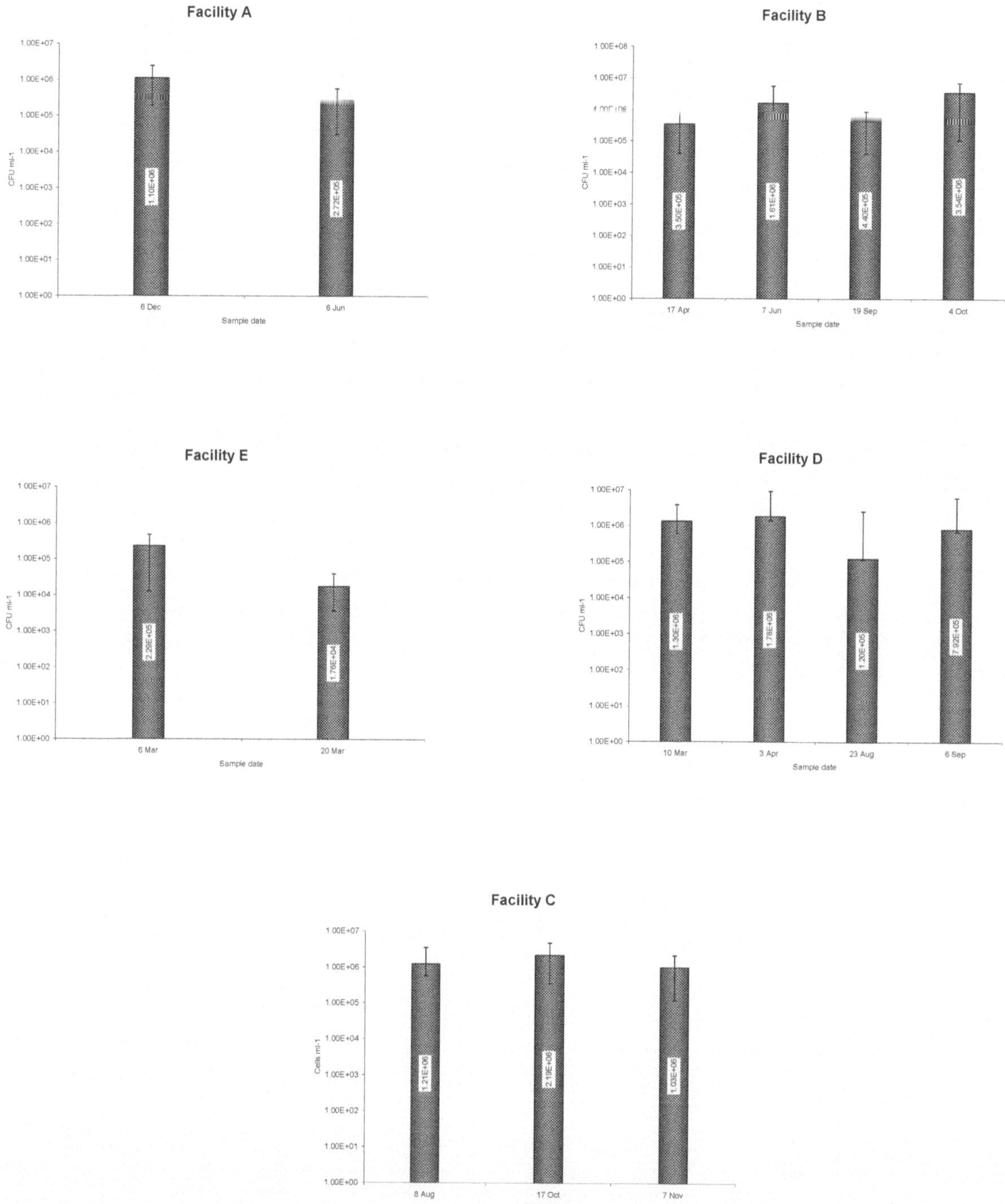

Figure 5-6. TCB Abundances for Effluent Samples from Facilities A-E.

In an effort to assess the applicability of TCB values to determine disinfection efficacy, $\log_{10}$ reduction values were calculated for each sample on days 2, 5, and 14 for chlorine and UV (Figure 5-12). The general trends at all facilities, except for Facility C (Fig. 5-12e), show a decrease in $\log_{10}$ reduction values, regardless of the disinfectant, with increasing incubation time. Tampa had only one of six samples (HOCl, August 8, 2001) that demonstrated a substantial decrease in $\log_{10}$ reduction values over the 14-day incubation period. When comparing the average net decreases in $\log_{10}$ reduction values after 14 days of incubation, each facility appeared to have a characteristic response (Table 5-3), though there was a general trend of greater decreases in $\log_{10}$ reduction values associated with UV irradiated TCB (Table 5-3).

Comparison of $\log_{10}$ reduction values for TCB after 14 days of incubation (Figure 5-12a) and FC (Figure 5-3a) from Facility A (Figure 5-13a) and Facility E (Figure 5-13c, compiled from Figure 5-3c and 5-12c) did not show a dominant trend. (Note: There were too few data points for effluent samples from Facilities A and E to perform a test of data normality.) There were no statistically significant differences between FC, Ent, and TCB $\log_{10}$ reduction values regardless of which disinfectant was used for effluent samples from Facility C (Figure 5-13, compiled from Figures 5-3, 5-5 and 5-12) ($p=0.511$) and Facility C (Figure 5-13, compiled from Figures 5-3, 5-5, and 5-12) ($p=0.407$) facilities.

However this was not the case for samples from Facility B. The $\log_{10}$ reduction values for TCB (Figure 5-13, compiled from Figures 5-3 and 5-12) following chlorine and UV disinfection were significantly different ($p=0.001$). However, TCB values following UV disinfection were equivalent ($p> 0.100$) to FC values following disinfection with chlorine and UV. TCB values from chlorinated samples were significantly different from FC values from chlorinated ($p=0.012$) and UV irradiated ($p=0.031$) samples. Using average $\log_{10}$ reduction values from each category as a general estimate of the magnitude of the significant difference, the TCB chlorinated samples were approximately three-fold lower than the chlorinated and UV irradiated values and approximately two-fold lower than UV irradiated TCB samples.

Collectively, these results indicate that what has been shown for the FC and Ent is also applicable to TCB within the context of this study. In particular, chlorine and UV radiation were similar in regard to their effectiveness for inactivation of FC, Ent, and TCB for the bench-scale conditions of exposure. The data also illustrate the effects of chemical and physical conditions that change within each facility with regard to disinfection efficacy. Moreover, it is important to reiterate that these conditions were sufficient to allow (bench-scale) disinfected samples to comply with facility-specific discharge regulations based on indicator bacteria.

12/6/00

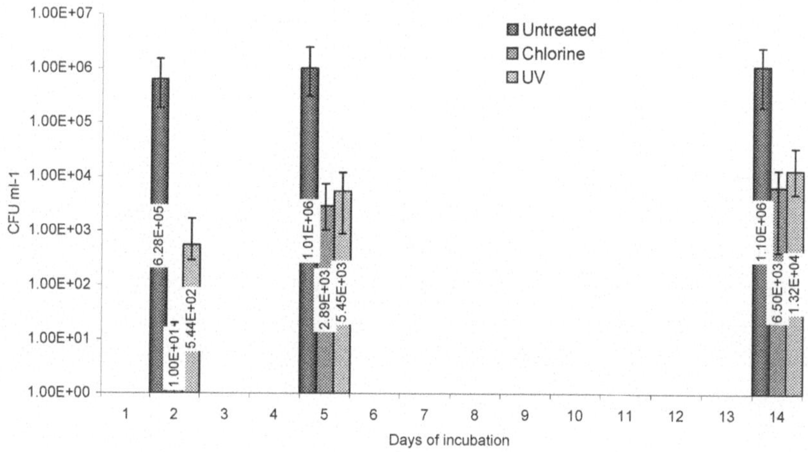

6/26/01

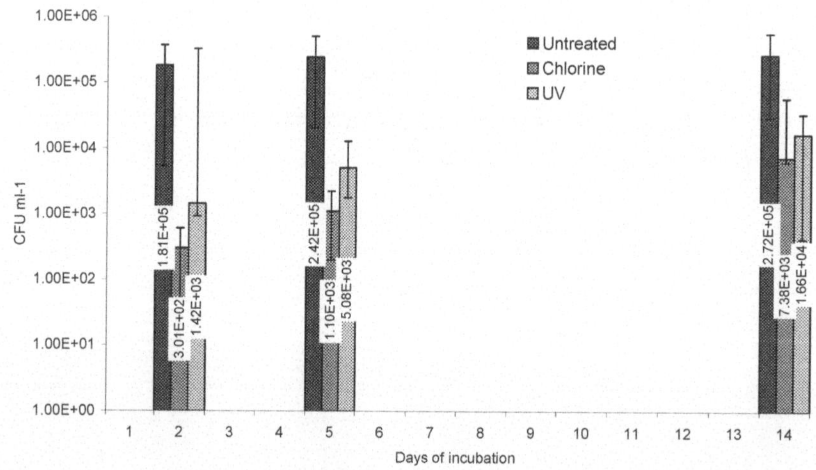

Figure 5-7. TCB Abundances in Facility A Samples Following Disinfection.

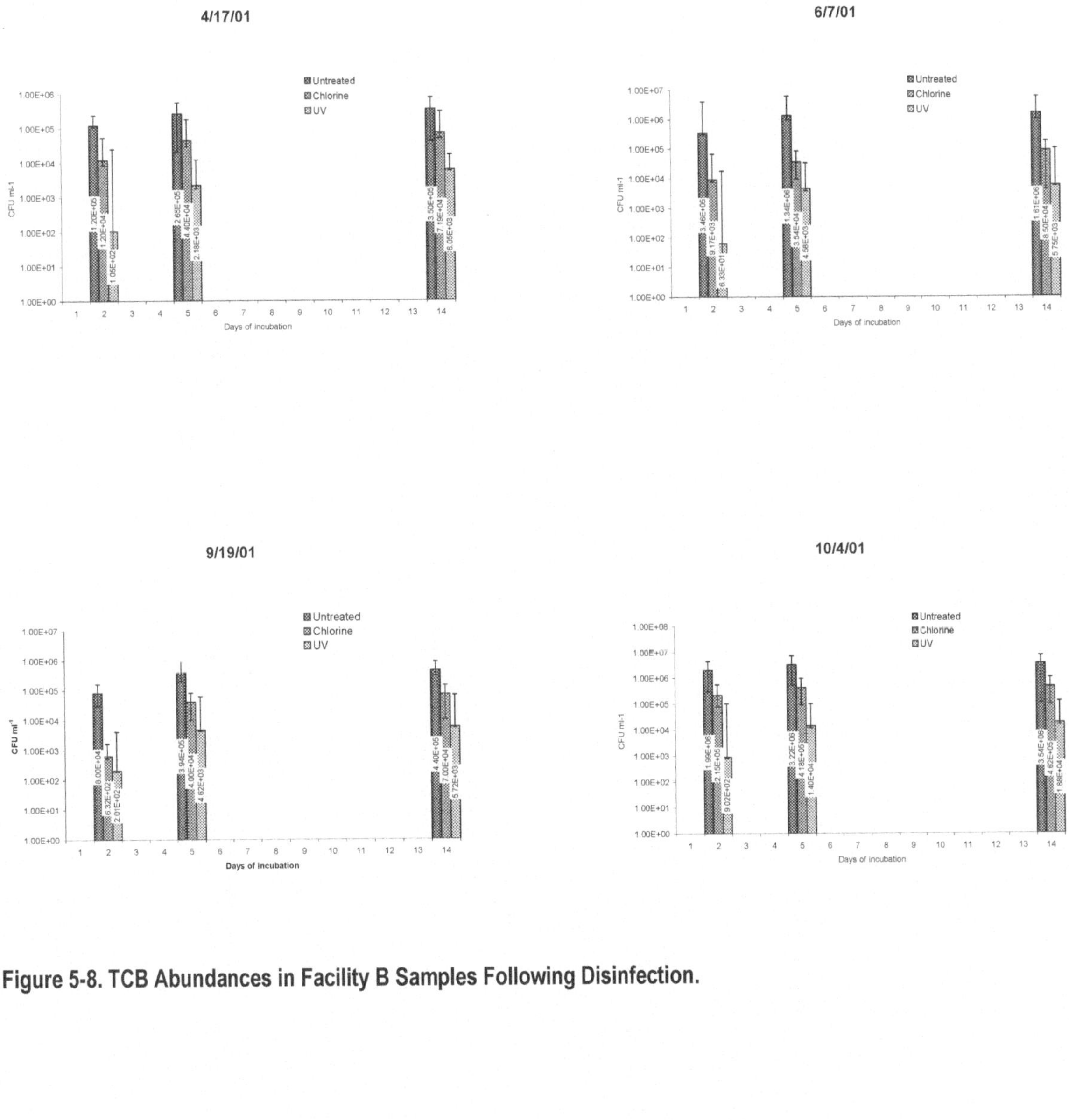

Figure 5-8. TCB Abundances in Facility B Samples Following Disinfection.

3/6/02

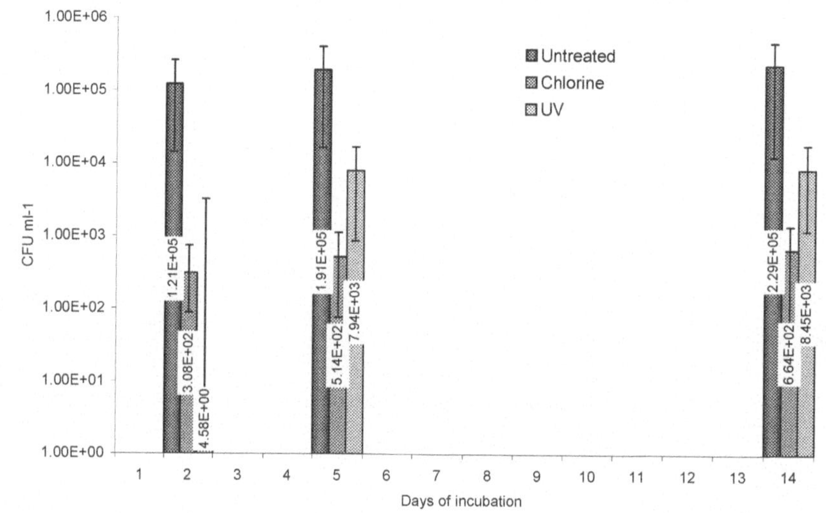

3/20/02

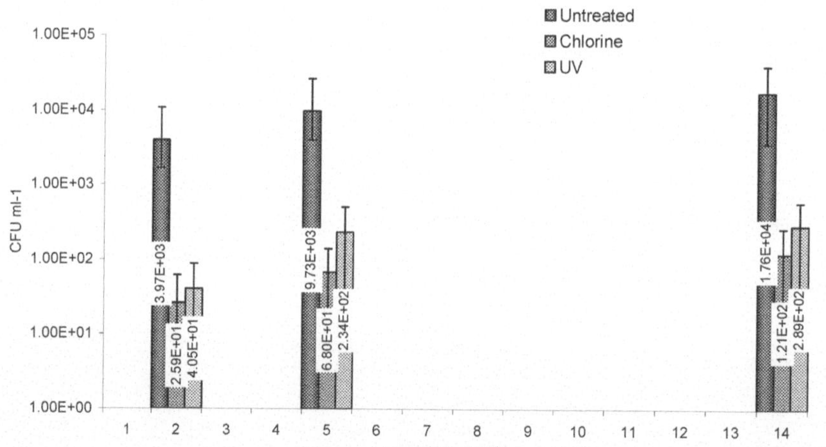

Figure 5-9. TCB Abundances in Facility C Samples Following Disinfection.

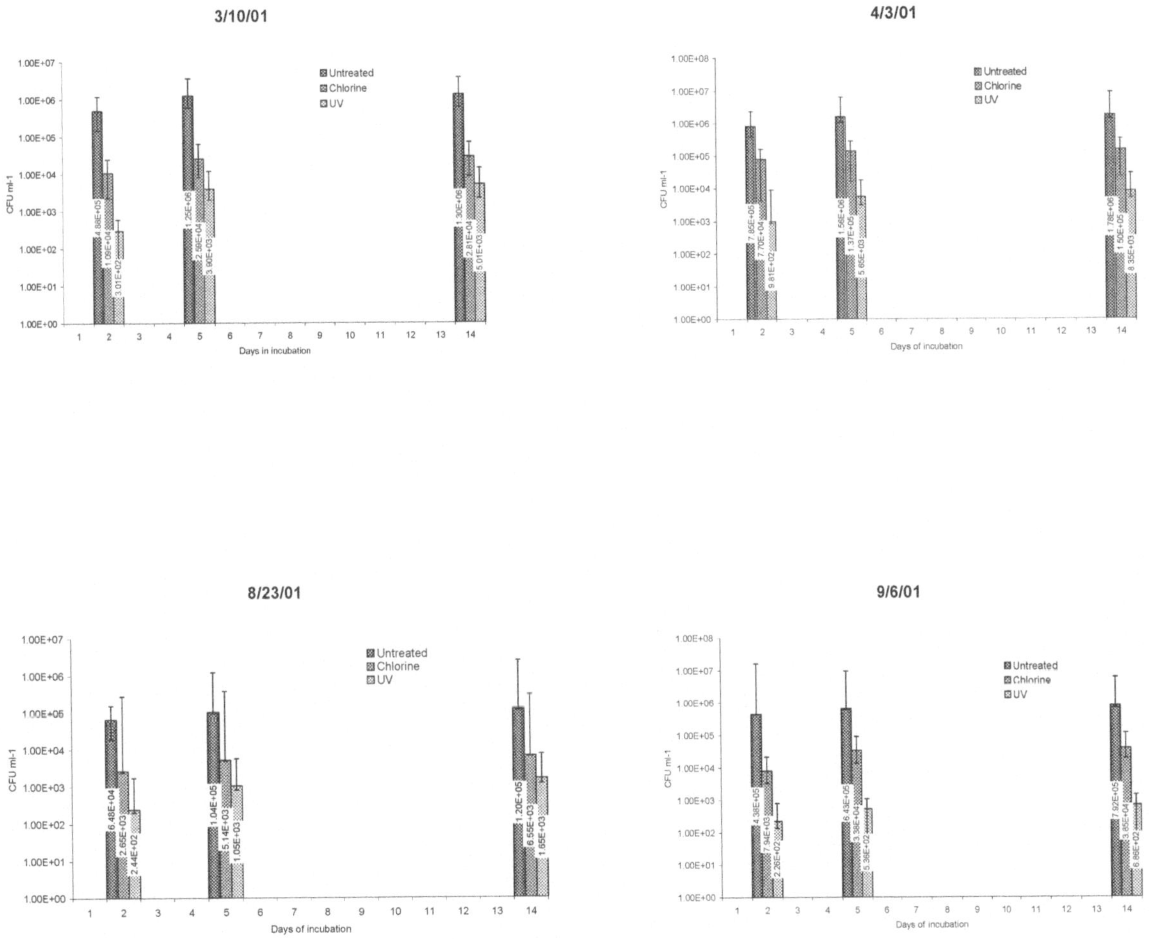

Figure 5-10. TCB Abundances in Facility D Samples Following Disinfection.

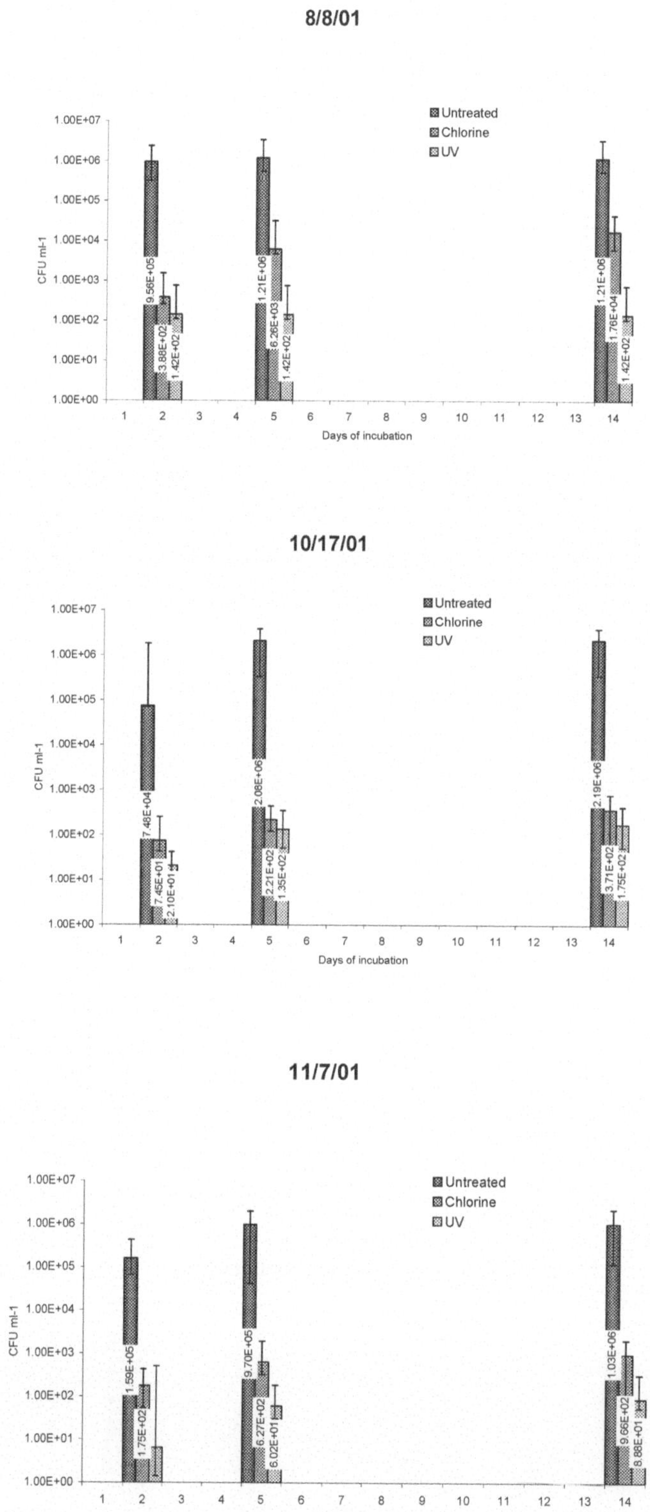

Figure 5-11. TCB Abundances Facility C Samples Following Disinfection.

WERF

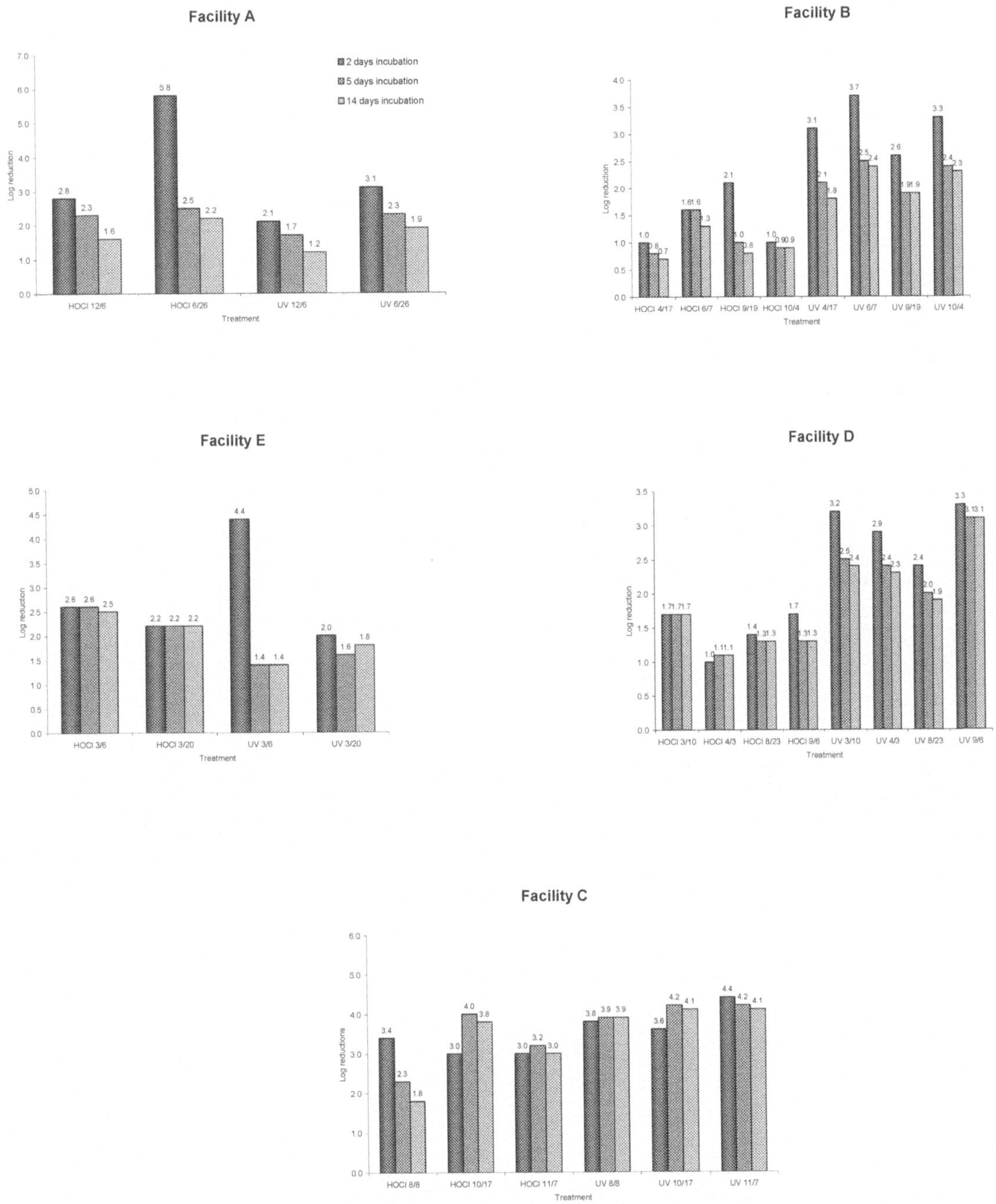

Figure 5-12. Comparisons in $\log_{10}$ Reductions of TCB Between Chlorine and UV for Samples Collected from Facilities A-E.

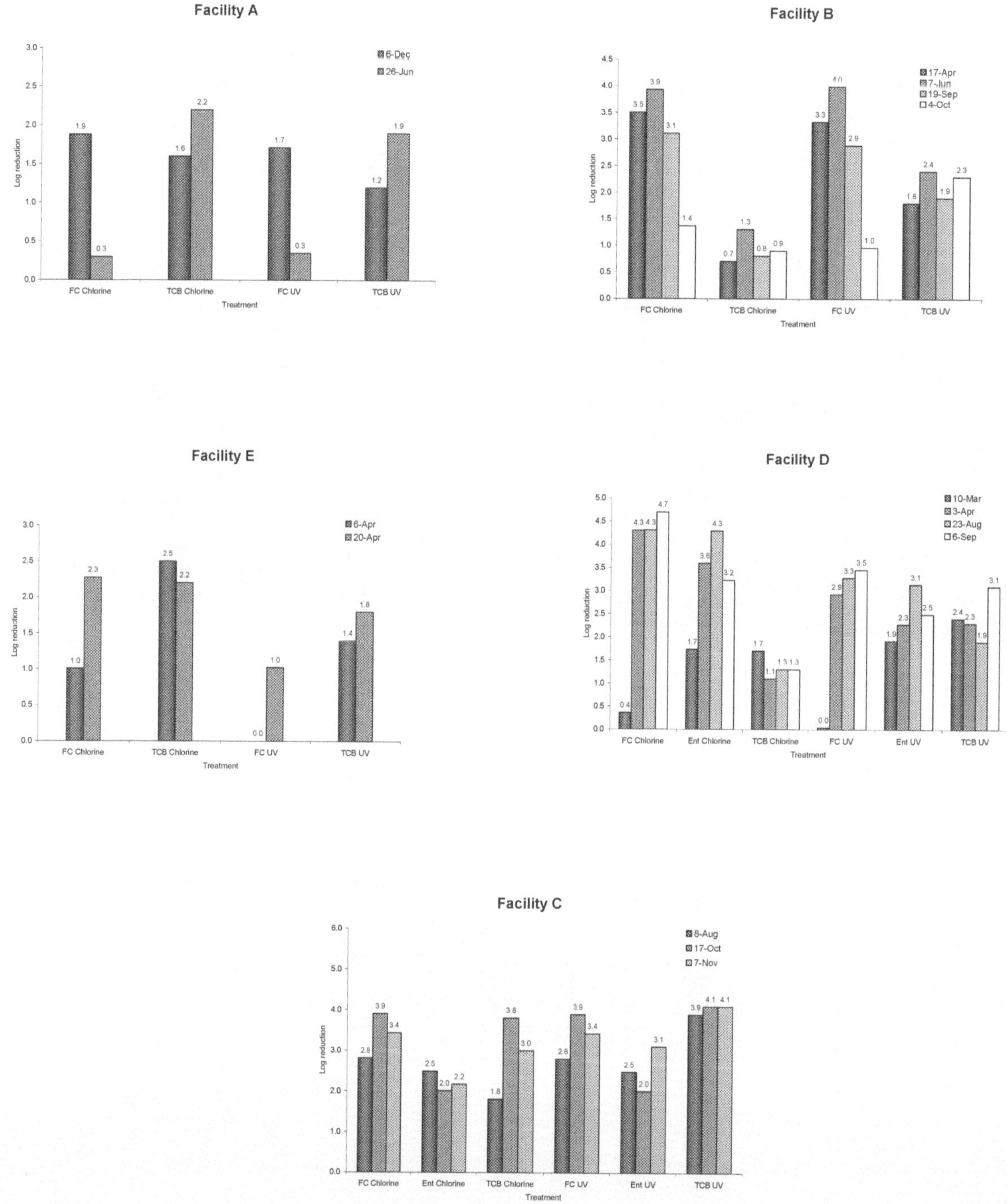

Figure 5-13. Comparisons Between log10 reductions of FC and TCB Following Disinfection by Chlorine and UV for Effluent Samples Collected from Facilities A-E.

5.4.5 The Effects of Oxyrase on Recovering Wastewater Bacteria Following Exposure to Chlorine or UV Radiation

All data for FC, Ent, and TCB, plated with and without exposure to Oxyrase, were pooled per treatment within each group. Each group was then compared to determine if Oxyrase improved culture-based recovery of the respective bacterial groups. There was no significant difference between the FC ($p=0.576$), Ent ($p=0.821$), and TCB ($p=0.853$) recovery values. These results indicate that the presence of Oxyrase did not significantly improve the recovery of oxidatively stressed bacteria and/or there was no appreciable production of reactive oxygen species on the filter membranes and media surfaces.

5.4.6 The Effects of Disinfection on Membrane Potential

Assessments of Reductions in Membrane Potentials Using Rh123 Following Disinfection

Bacterial viability has been defined several ways, including the ability to replicate on or in an appropriate medium (i.e., culturable), having intracellular enzyme activity that is directly related to a required metabolic pathway (e.g., succinate dehydrogenase) and possessing a membrane potential (McFeters et al., 1999). Though each approach to assessing physiological activity in bacteria has its proponents, membrane potential was included in this investigation for the following reasons: a) a membrane potential is the most important aspect of a bacterium's metabolism as it is the central component of the chemiosmotic system that is responsible for generating energy equivalents, such as adenosine triphosphate (ATP) (Saier, 1997); b) ATP reserves in an active bacterium will sustain a bacterium for only minutes following the collapse of the membrane potential (Barrett et al., 1989; Kinniment and Wimpenny, 1992); c) chlorination negatively affects membrane potential in bacteria (Barrett ct al., 1989; Camper and McFeters, 1979); and d) the presence or absence of the membrane potential in a bacterial cell is relatively easy to assess using fluorescent microscopy and fluorescent probes that are distributed across or excluded from crossing the membrane by cells with a membrane potential (Kaprelyants and Kell, 1992; Lisle et al., 1999).

The abundance of physiologically active bacteria, based upon the presence of a membrane potential, was determined by differential microscopic counts. The number of cells having a detectable membrane potential due to the intracellular concentration and retention of Rh123 was expressed as a ratio of the TDC. In this study, physiological activity of bacteria from each facility was assessed following disinfection with chlorine and UV radiation (Figure 5-14). The general trend was that chlorination appeared to negatively affect membrane potential more than UV irradiation. There are several samples, however, where membrane potentials were affected similarly by exposure to chlorine or UV radiation. In only one sample did UV irradiation decrease the percentage of cells with a membrane potential to a value lower than the chlorination sample (Figure 5-14D, September 6).

Whereas both disinfectants were relatively efficient at inactivating FC, Ent, and TCB as defined by culture-based methods, $\log_{10}$ reductions of similar magnitudes were not evident in the membrane potential data sets (Figure 5-15). Log_{10} reduction values ranged from 0.2 to 1.1 following chlorine disinfection and no reduction in cells with a membrane potential to 1.3 $\log_{10}$ reduction for UV disinfection. Comparison of pooled $\log_{10}$ reduction values from Facility B and Facility D indicated there were no significant differences between the disinfectants in inactivating cells as determined by the presence of a membrane potential.

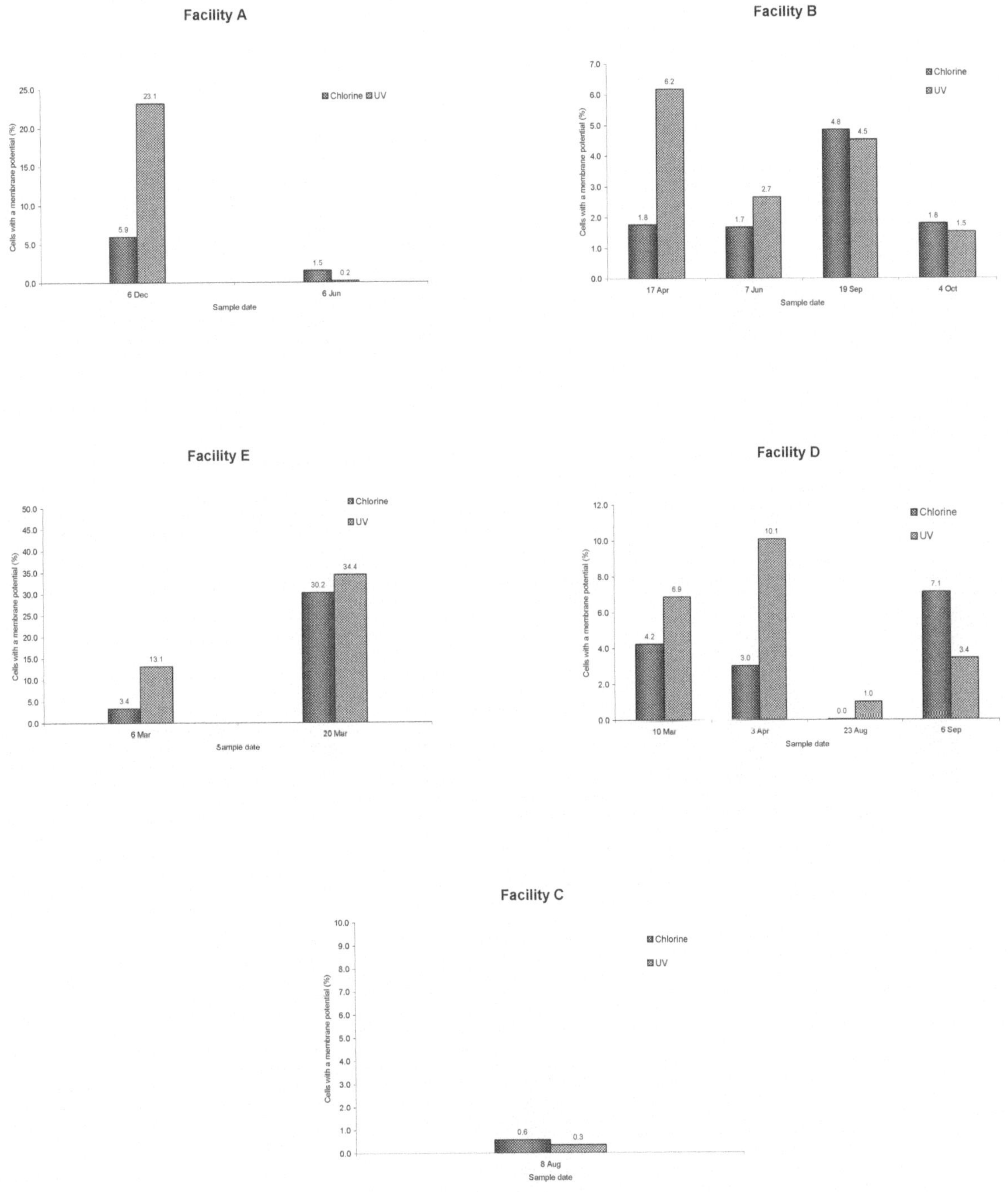

Figure 5-14. Assessment of the Presence of Membrane Potential in the Total Bacterial Population Following Disinfection for Effluent Samples from Facilities A-E.

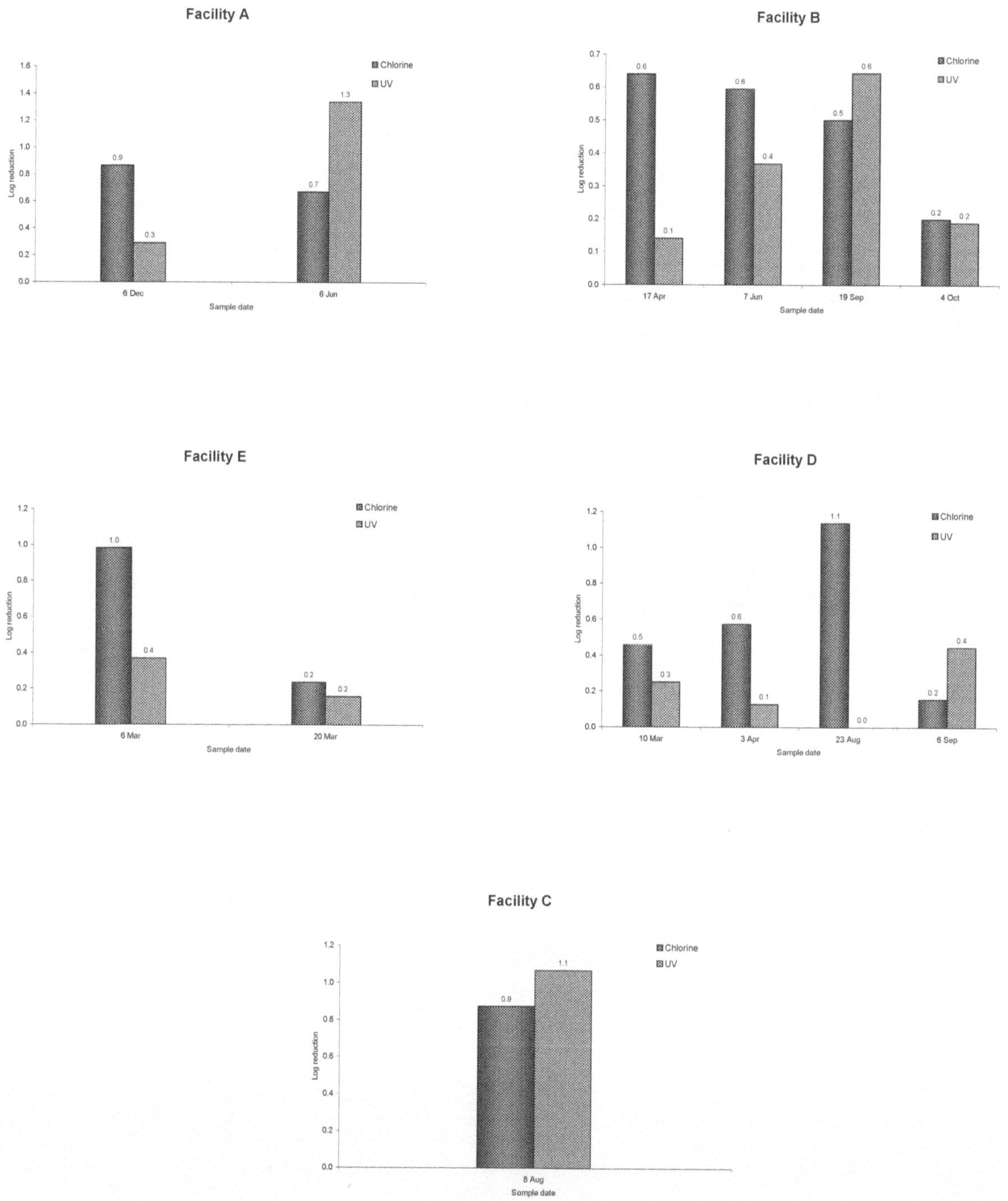

Figure 5-15. Log_{10} Reductions in Membrane Potential Following Disinfection for Effluent Samples Collected from Facilities A-E.

Assessments of Cells that Do Not Have Membrane Potentials Using DiBAC Following Disinfection In addition to assessing bacterial physiology as a measure of cells with membrane potential, a complementary fluorescent probe, $DiBAC_4$ (APHA, 1998) (aka, DiBAC) that is actively excluded from cells with membrane potentials was also applied (Lisle et al., 1999; Nebe-von Caron et al., 1998). The use of DiBAC and Rh123 should have provided an assessment of the respective absence and presence of membrane potentials in all cells that were untreated and exposed to the disinfectants. Ideally, the sum of both assays should have equaled the TDC. In only one sample (Figure 5-16a, December 6) was there a significant increase in the number of cells without a membrane potential following disinfection with chlorine and UV and when compared to the Rh123 values for the same sample (Figure 5-14a). Only the UV data accounted for a value (103.1%) approaching that of the respective TDC value, while the data from chlorination experiments accounted for only 88.9% of the TDC. The other facilities had similar values for cells without a membrane potential following exposure to chlorine and UV disinfection, but accounted for substantially lower percentages of the respective TDC (Figure 5-16b-e). From these samples, the combined Rh123 and DiBAC data for chlorine-based disinfection ranged from 1.1% (Figures 5-14e and 5-16e) to 16% (Figures 5-14b and 5-16b, September 19) of the TDC, while UV irradiation data ranged from 1.2% (Figures 5-14d and 5-16d, April 3) to 14% (Figures 5-14e and 5-16e, March 20).

Since DiBAC labels cells without a membrane potential, there should be a concomitant $\log_{10}$ increase in the abundance of $DiBAC^+$ cells equal to the $\log_{10}$ decrease of $Rh123^+$ cells (i.e., cells with a membrane potential). Figure 5-17 illustrates the $\log_{10}$ increases in the abundances of cells without a membrane potential (i.e., $DiBAC^+$ cells) for each facility and the respective sample dates. The $\log_{10}$ increase values for the chlorination assay ranged from no $\log_{10}$ increase in $DiBAC^+$ cells to 2.0. The ranges of $\log_{10}$ increases following UV irradiation were similar, ranging from no $\log_{10}$ increase in $DiBAC^+$ cells to 1.4. Comparison of pooled $\log_{10}$ reduction values from the Lafayette (Figure 5-17b; $p=0.869$) and St. Petersburg (Figure 5-17d; $p=0.129$) facilities indicated there were no significant differences between the disinfectants in inactivating cells as determined by the absence of a membrane potential.

Also, considering that $Rh123^+$ cells decrease and $DiBAC^+$ cells increase following exposure to a disinfectant, a negative correlation would be predictable. Using pooled $\log_{10}$ reduction values from Figures 5-15 (Rh123) and 5-17 (DiBAC) for chlorine and UV disinfected cells, there were no significant correlations, negative or positive, ($p \geq 0.10$) between Rh123 and DiBAC data sets.

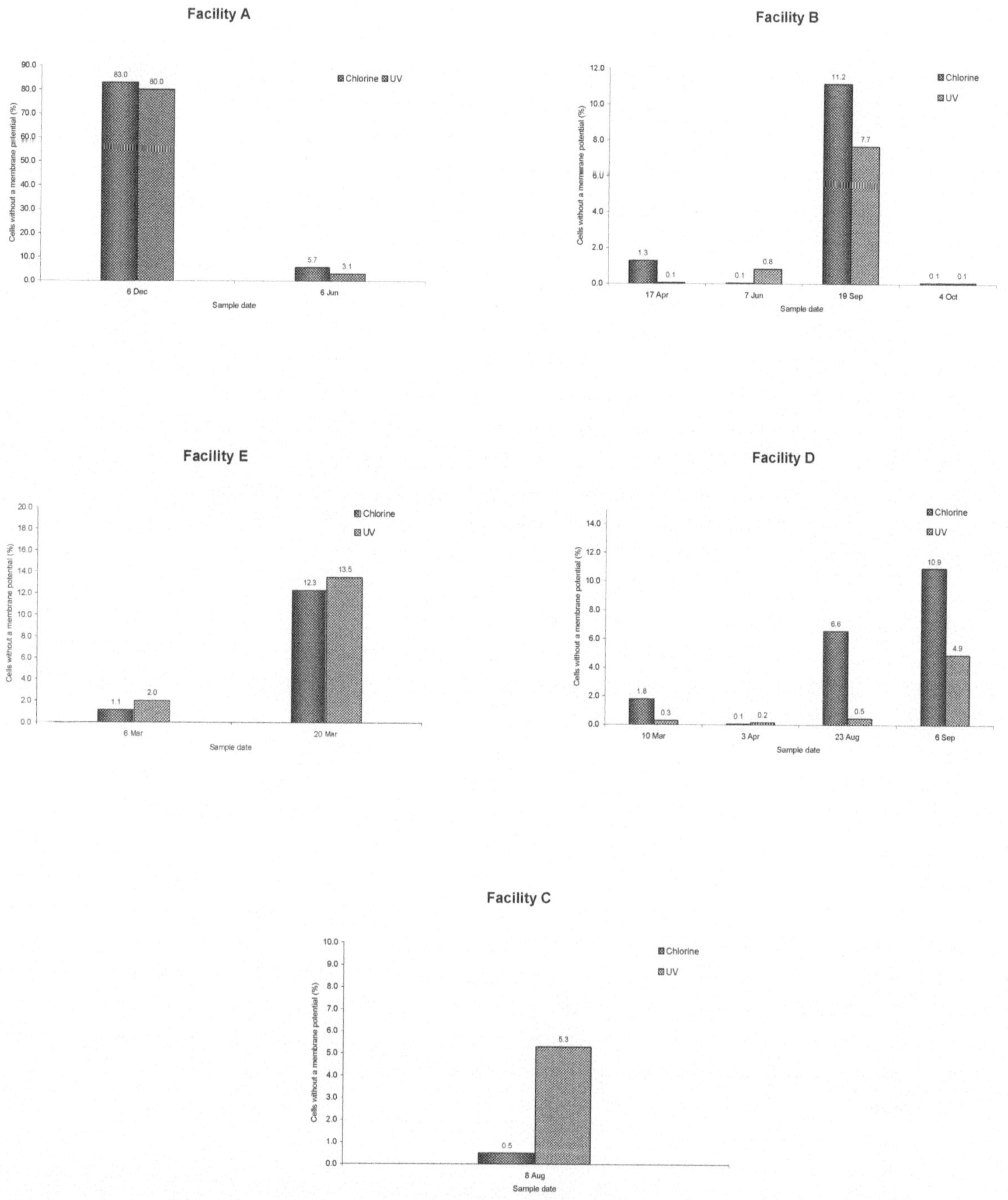

Figure 5-16. Assessment of the Absence of Membrane Potential in the Total Bacterial Population Following Disinfection for Effluent Samples Collected from Facilities A-E.

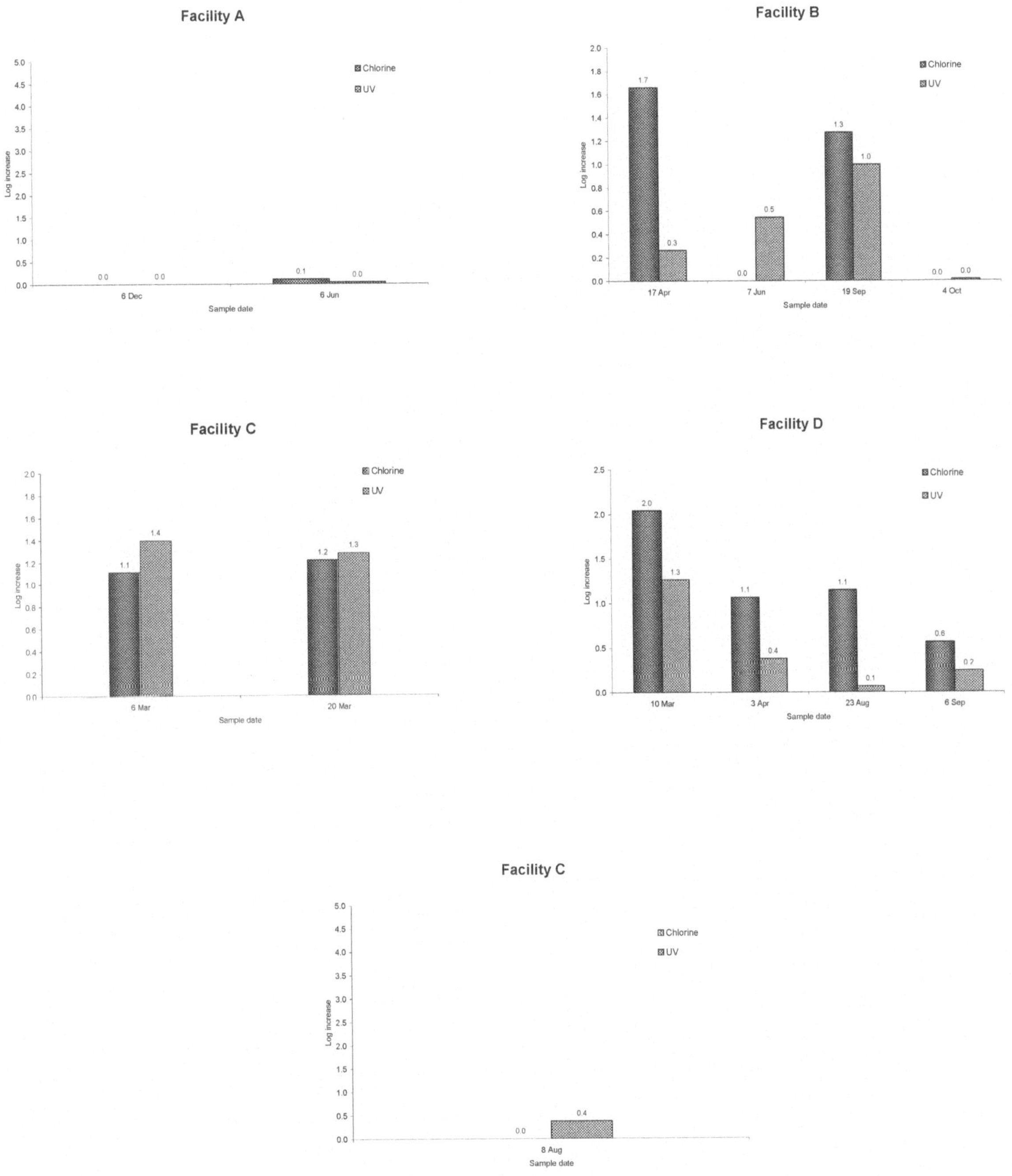

Figure 5-17. Log$_{10}$ Increases in the Absence of Membrane Potential Following Disinfection for Effluent Samples Collected from Facilities A-E.

Collectively, these data suggest that membrane potential assays are insufficiently sensitive for use in characterizing inactivation of wastewater bacteria by chlorination or UV radiation. Several factors influence the efficient labeling of bacterial cells with Rh123 and DiBAC (McFeters et al., 1999; Shapiro, 1990). These include a) physiological status of the cell; b) type of bacterial membrane (i.e., Gram + or -); c) probe concentration; d) labeling time; and e) presence of bacterial transport systems that actively pump out the probe. Additionally, bacterial membranes that are sub-lethally stressed or compromised may not efficiently interact with the probes though a membrane potential exists, leading to the distribution of the probe into the cell, but at a concentration that is not detectable using the microscopic methods employed in this study. There is some evidence that the majority of samples contained significant percentages of injured bacteria upon receipt from sample collection sites (Table 5-4).

Table 5-4. Summary of Bacterial Injury Measurements.

Facility Identifier	Sample Collection Date	Injury at Time of Receipt
A	12/6/2000	30.8
B	4/17/2001	77.8
	6/7/2001	10.0
	9/19/2001	33.3
	10/4/2001	1.0
E	3/6/2002	94.4
D	3/10/2001	35.5
	4/3/2001	8.6
	8/23/2001	1.0
	9/6/2001	41.1
C	8/8/2001	>95.0
	10/17/2001	90.4
	11/7/2001	93.8

The injury assay used in this study was chosen as a tool to assess membrane injury in the indigenous bacterial populations of the respective samples prior to and following disinfection. The assay provided very little data on post disinfection samples, as the data were extremely variable within and between samples. These data suggest that the collection, storage, and shipping conditions for most of the samples significantly and negatively compromised the membranes of the bacterial populations. The inhibitory effects of sample storage on culturability of bacterial populations are well established (Cox and Claiborne, 1949; McDaniels and Bordner,

1983; Zobell, 1943). These high percentages of injured bacteria in the samples prior to disinfection predispose the bacteria to the actions of the disinfectants, especially chlorine. This situation confounds the Rh123 and DiBAC data and makes their application to assessing the efficacy of disinfectants in this study questionable. Preliminary research using Rh123 and DiBAC in wastewater treatment facility effluent from the Houston, Texas area were successful (data not shown). The major difference in the two types of samples is that the preliminary samples were processed immediately for labeling with Rh123 and DiBAC. Therefore, the use of fluorescent labels and probes to assess bacterial physiological activity in wastewater samples prior to and following exposure to disinfectants remains as a potentially relevant option to culture-based techniques. However, their application may be best conducted on site or after a very short storage and transportation period, avoiding the inhibitory effects of sample storage and shipment.

5.5 Conclusions

The data from the culture- and non-culture based bacterial assays reveal several interesting trends regarding bacterial responses to common disinfectants (chlorine and UV radiation), as well as the methods used to characterize them.

Among the various bacterial indicator groups, the TCB group appears to be roughly equivalent to the FCs and Ent in assessing disinfection efficacies of chlorine and UV radiation in treated wastewater. However, and important disadvantage of using total culturable bacteria for assessing disinfection efficacy is the assay requires as much as 14 days incubation to provide reliable data.

In general, chlorination and UV irradiation were similarly efficient at inactivating FCs, Ent, and TCB in the types of wastewater samples analyzed in this study. However, there were samples where chemical and physical components in and of the sample reduced each disinfectant's antibacterial efficacy.

The abundance of indicator and non-indicator bacterial groups in a wastewater stream changes dramatically over short periods. This dynamic situation suggests that an optimized disinfection system should focus on the highest abundance values for the selected indicator group. A fixed (constant) dose of disinfectant applied to a wastewater stream during a 24-hour period will not be sufficient to inactivate the target bacterial group if that dose has not been adjusted to maximum abundance data. This approach will ensure adequate disinfection to meet the effluent regulations for effluents.

Both disinfectants negatively influenced the growth of FC, Ent, and TCB. Increased incubation periods at lower temperatures increased recovery of disinfectant injured TCB, providing a more reliable assessment of disinfection efficacy.

Non-culture based methods for assessing disinfection efficacies indicated that both disinfectants reduced physiological activity to a similar degree. However, these data should be considered conservative indicators of how the disinfectants affect bacterial physiology as sample storage and shipping conditions significantly injured a large fraction of bacteria in most of the samples analyzed.

CHAPTER 6.0

Effects of UV Irradiation and Chlorination/Dechlorination On Indigenous Bacteriophage

6.1 Background

Traditionally, assessments of antimicrobial efficacy in disinfection operations used for treatment of municipal wastewater have been based on measurements of the concentration of viable indicator bacteria. While these organisms satisfy some of the basic requirements of indicator organisms, several important shortcomings of their application for this purpose have been identified (see Chapters 3.0 and 5.0). Among the most important of these limitations are the relative ease with which most bacterial indicators are inactivated by common disinfectants and the fact that enteric viruses often represent the most serious risk to human health.

Unfortunately, the assays used to assess viability (or infectivity) among human enteric viruses are time-consuming and expensive. In most situations, it is not practical to monitor for human enteric viruses. However, several indigenous phage have been identified that are structurally or otherwise similar to human viruses. Assays of phage viability (infectivity) are comparatively easy to conduct. Therefore, a series of experiments was conducted to assess the effects of common wastewater disinfectants on the concentrations of viablc (infective) indigenous phage.

6.2 Materials and Methods

Non-disinfected secondary effluent samples were collected from Facilities A-E. After receipt, effluent samples were split to allow analysis of viable phage concentration in non-disinfected subsamples, as well as sub-samples that were subjected to chlorination/dechlorination or UV irradiation using the batch exposure protocols described previously (briefly summarized below). To reiterate, these disinfection exposure scenarios were demonstrated to be effective for treating effluent samples from these facilities so that compliance with discharge regulations relevant to disinfection (i.e., viable indicator bacteria concentration and residual chlorine concentration) could be accomplished.

Coliphage analysis was performed by the double agar overlay method (Adams, 1959). Total coliphages (somatic and F-specific) were assayed on tryptic soy agar using a log-phase host culture of *E. coli* C-3000 (ATCC #15597). F-specific coliphages (F+ phages) were assayed using a log-phase culture of *E. coli* F_{amp}-HS (pFamp)R (ATCC #700891) on tryptic soy agar augmented with streptomycin/ampicilin as specified by U.S. EPA (2001).

Individual phage plaques surviving either 40 minutes contact time with an initial chlorine concentration of 2.0 mg/L as Cl_2 (added to a well-mixed batch reactor in the form of an aqueous NaOCl solution) or a UV dose of 20 mJ/cm^2 (as delivered by a collimated beam) were harvested and stored in 0.5 mL aliquots of PBS at 4°C, then regrown to high titer. Bacteriophage isolated post-disinfection were further characterized by their nucleic acid content. Bacteriophage containing RNA were differentiated from those containing DNA by suppressing growth of isolates in the presence of 100 µg/mL of RNAse (bovine pancreas, type I-A, Sigma). Isolates that did not produce plaques in the presence of RNAse were classified as RNA phage. Bacteriophages that were able to form plaques in the presence of RNAse were classified as DNA phage.

6.3 Results and Discussion

The concentration of indigenous bacteriophage in effluent samples from the five wastewater treatment facilities varied considerably, with the highest phage concentrations isolated from Facility B. The phage population for this facility was comprised of somatic and F+ specific phage, with Facility B possessing the highest concentration of F+ specific phages of all the facilities examined. In decreasing order of initial phage concentration, the facilities were ranked as: B > A > D ≈ E > D. The concentrations of phage in samples from Facilities D and E were similar. The samples containing the highest concentration of phage surviving either chlorine or UV disinfection appeared to reflect the ranking of the facilities with regard to initial phage concentration.

To characterize the effects of particulate matter on protection of bacteriophage from UV disinfection, several of the secondary wastewater samples were prefiltered through 0.22 µm filters prior to performing the UV dose response assays (Figure 6-1).

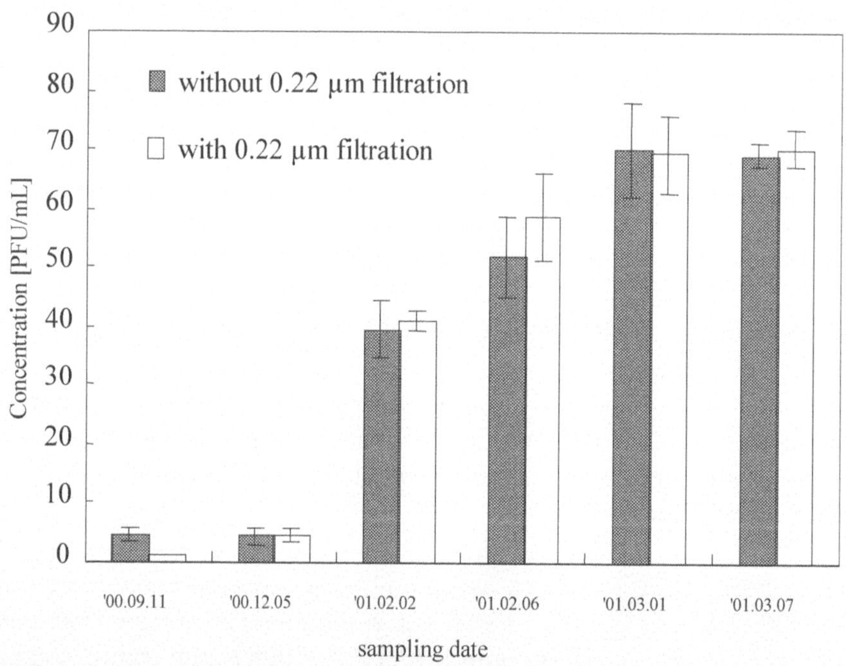

Figure 6-1. Effect of 0.22 µm Filtration on Phage Concentration.

The concentrations of bacteriophage in samples before and after filtration were similar, with the exception of the September 11, 2000, sample which had extremely low phage concentrations such that changes in concentration could not be meaningfully established. Accordingly, the physical aspect of particle attachment of bacteriophage was considered to be negligible. However, the effect of particles whose size was <0.22 µm was not determined. The smallest phage categorized is approximately 20 nm; therefore, a clump consisting of tens of phages would be less than the pore size of a 22 µm filter.

Figure 6-2 depicts the UV inactivation of phage in 0.22 µm filtered and unfiltered secondary effluent samples. Filtration through a 0.22 µm membrane filter did not significantly affect the concentration of viable phage in irradiated or non-irradiated samples. Therefore, the particles retained on the 0.22 µm filter did not appear to provide substantial protection to the phage in this water sample, nor were a significant fraction of the phage embedded in these particles.

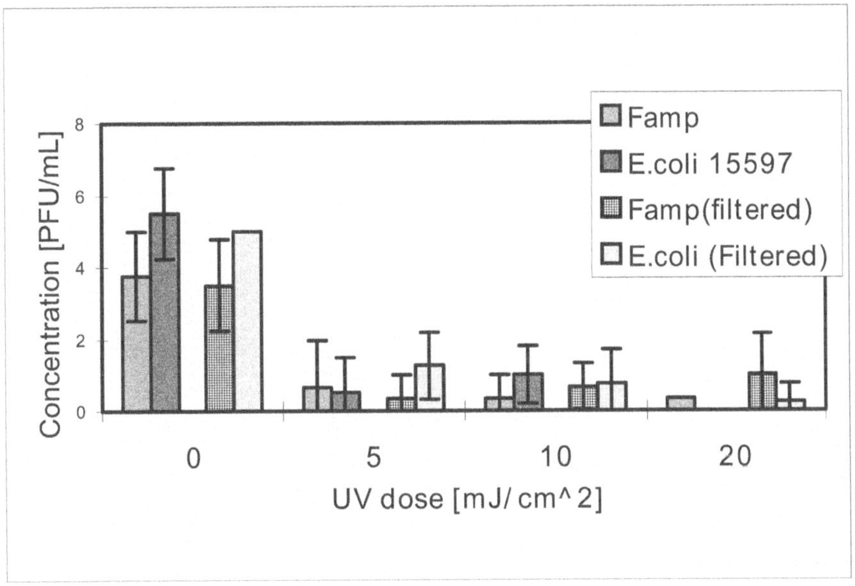

Figure 6-2. UV Inactivation of Indigenous Phage with and without 0.22 µm Filtration. Sample Collected December 5, 2000, from Facility A.

6.31 Facility A Samples

The initial concentration of coliphage and F+ phage varied dramatically in samples collected from Facility A from 2000 through 2001 (Figures 6-3 through 6-8). In samples collected during 2000, the phage concentration was relatively low and ranged from a low of 3 pfu/mL to a high of 13 pfu/mL (Figures 6-3 through 6-5). In samples collected during 2001, however, the bacteriophage concentration increased dramatically ranging from a low of 40 pfu/mL to a high of 198 pfu/mL (Figures 6-6 through 6-8). Bacterial analysis performed by co-investigators on the same samples showed extremely high bacteria TDCs.

It is likely that the variation in phage and bacterial concentration among samples collected during the years 2000 and 2001 was attributable to seasonal effects. Examination of historical records from Facility A (and similar facilities located in temperate climates) indicates a general trend towards higher concentrations of bacteria during warm weather months. More specifically, it is common to observe increases in concentrations of indicator (coliform) bacteria of roughly an order of magnitude from samples collected in winter months to similar samples collected in summer months. The results of bacteriophage analysis presented above suggest that similar seasonal variation in phage concentration can be expected.

While initial work revealed that filtration with a 0.22 μm membrane did not appreciably alter the concentration of bacteriophage by removal of bacteria and other large particles in the samples evaluated, this work was performed on year 2000 samples, where phage counts as well as the total concentration of bacteria were low. Previous studies have reported increased survival of bacteriophage in samples through the formation of protective bacterial aggregates, within fecal matter, or inclusion into bacteria or higher organisms such as amoeba, nematodes and crustaceans (Chang et al., 1960; Chang, 1961; Tracy et al., 1966). The protection mechanisms of particle-associated organisms involve diffusion limitations inside aggregates, the reduction of the organisms' exposed surface, as well as disinfectant demand exerted by particles. In the case of UV disinfection, particle-associated protection is assumed to involve shielding of the bacteriophage by larger particles and organisms.

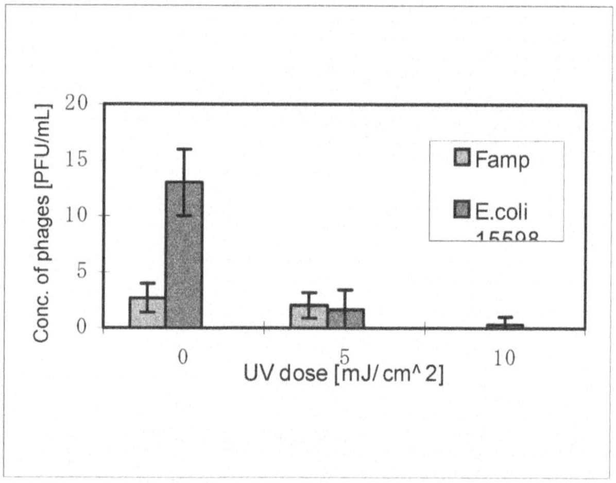

Figure 6-3. UV Inactivation of Indigenous Phage.
Sample collected from Facility A, October 11, 2000.

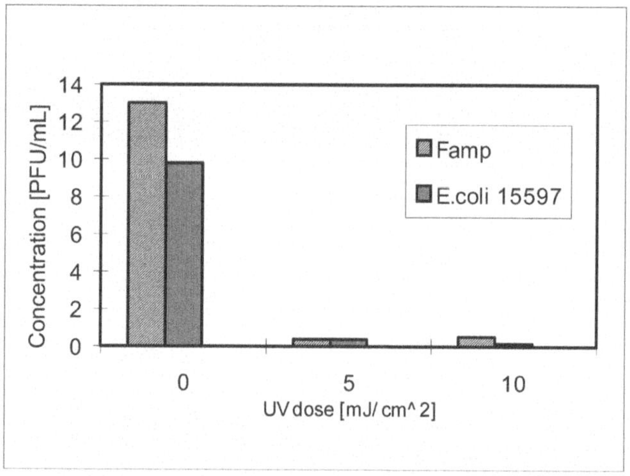

Figure 6-4. UV Inactivation of Indigenous Phage.
Sample collected from Facility A, November 9, 2000.

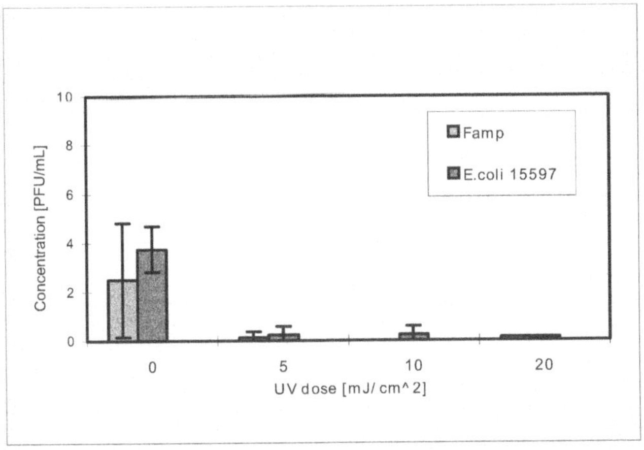

Figure 6-5. UV Inactivation of Indigenous Phage. Sample Collected from Facility A, November 28, 2000.

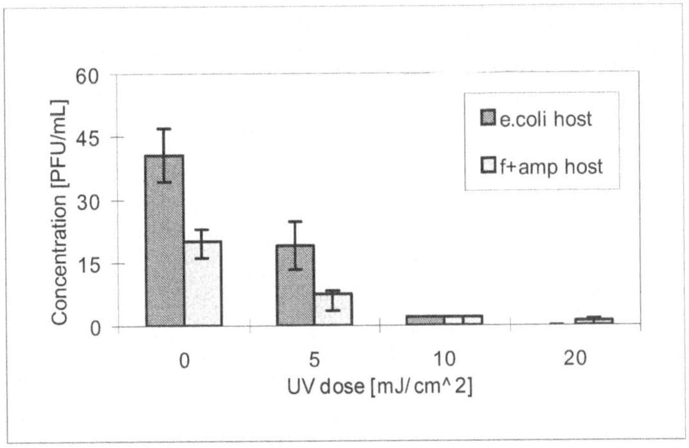

Figure 6-6. UV Inactivation of Indigenous phage. Sample Collected from Facility A, June 6, 2001.

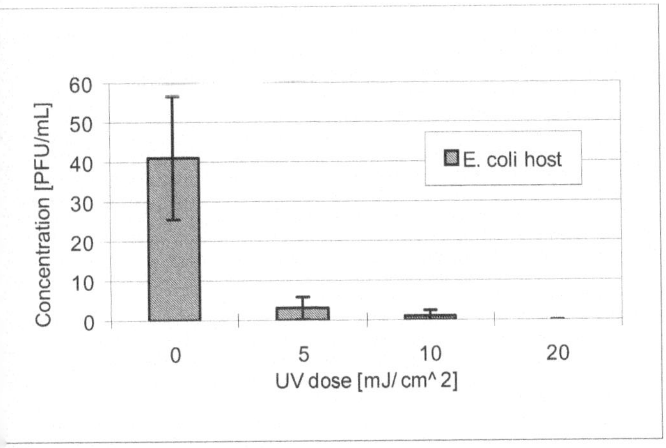

igure 6-7. UV Inactivation of Indigenous Phage. Sample ollected from Facility A, July 12, 2001.

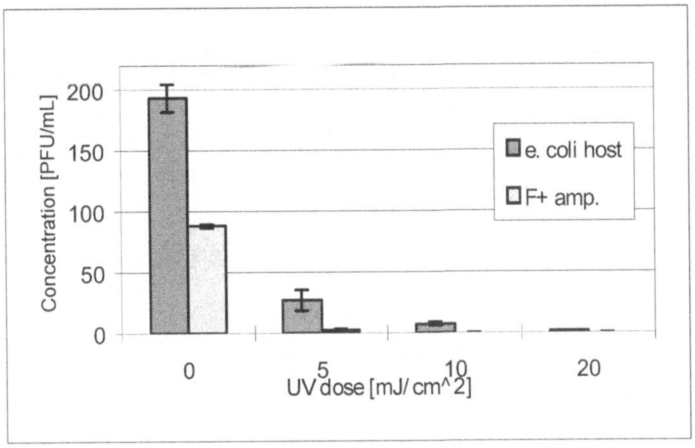

Figure 6-8. UV Inactivation of Indigenous Phage. Sample Collected from Facility A, September 19, 2001.

The UV sensitivity of the phages isolated from Facility A samples varied substantially (Figure 6-9). None of the bacteriophage that were regrown from plaques isolated from the Facility A samples survived a UV dose of 20 mJ/cm^2 when reevaluated. This suggests that the UV resistance of the indigenous bacteriophages was not due to genetic or physiological attributes of the phage isolates, but was due to physical factors within the wastewater matrix.

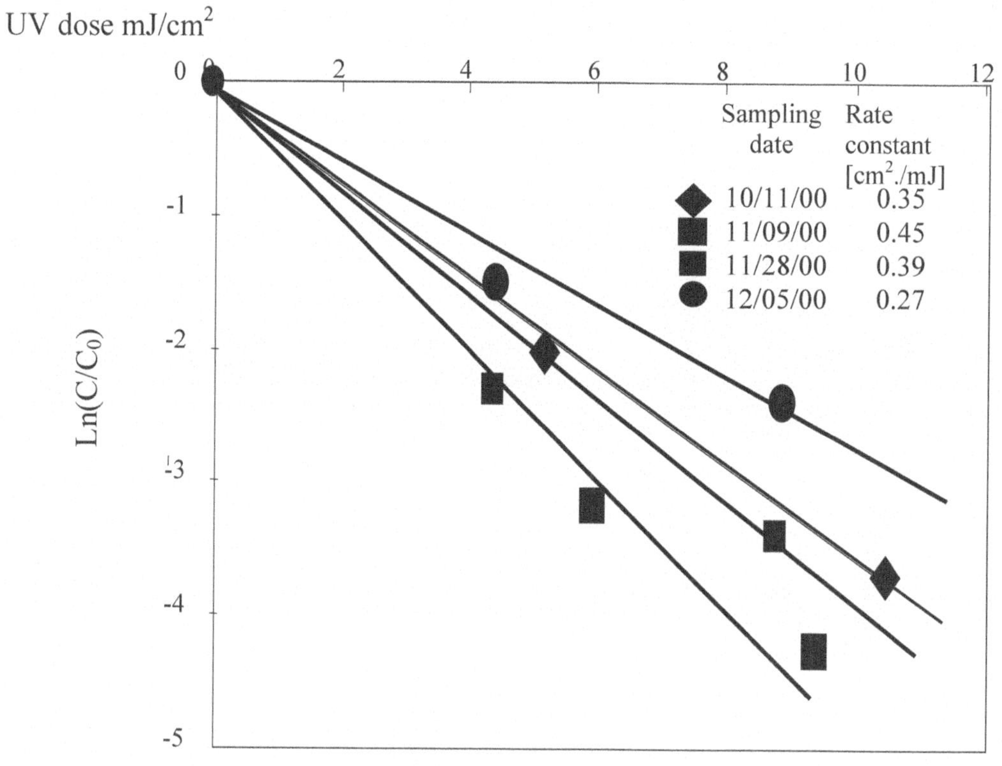

Figure 6-9. Indigenous Bacteriophage Variability in UV Sensitivity in Samples Collected from Facility A.

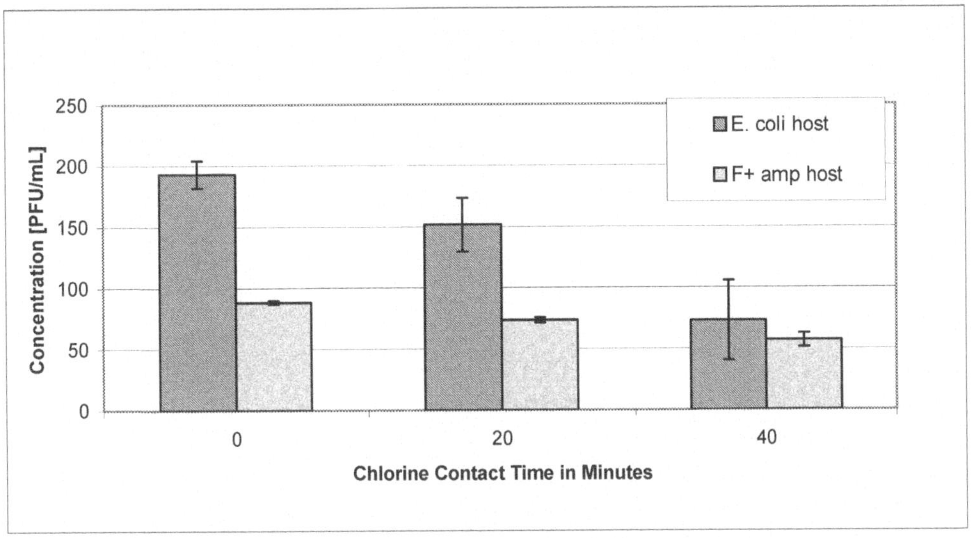

Figure 6-16. Chlorine Inactivation of Indigenous Phage Based on an Initial Chlorine Concentration of 2.0 mg/L (as Cl_2). Sample Collected from Facility A, September 19, 2001.

Table 6-1. Nucleic Acid Content of Post-disinfection Viable Phage in Facility A Samples.

Disinfectant Exposure Scenario	Host Strain	Nucleic Acid Content	
		RNA	DNA
40 min contact time; 2.0 mg/L as Cl_2	F+ amp	7	4
40 min contact time; 2.0 mg/L as Cl_2	*E. coli*	3	2
20 mJ/cm^2 UV	F+ amp	0	0
20 mJ/cm^2 UV	*E. coli*	0	3

6.3.2 Facility B Samples

Samples collected from Facility B showed a similar initial bacteriophage concentration to that of Facility A (Figures 6-17 through 6-20). Prefiltering the wastewater sample with a 0.22 μm filter prior to UV disinfection did not alter the phage inactivation (Figure 6-17). There were no bacteriophage detected in any of the Facility B samples, regardless of initial phage concentration, after a UV dose of 20 mJ/cm^2 (Figures 6-17 through 6-20). There were, however, a number of bacteriophage that survived 40 minutes of chlorine contact time, based on an initial chlorine concentration of 2.0 mg/L (as Cl_2) (Figures 6-21 through 6-23). Two of these samples contained initial coliphage concentrations >300 pfu/mL. While coliphage are found infrequently in human feces, they often are detected in high concentrations in wastewater treatment facility effluent prior to disinfection. This suggests that they may have been introduced to the WWTP from sources other than human feces, such as stormwater (Havelaar et al., 1986).

At the time of sample collection, Facility B operated as a conventional activated sludge facility, without nitrification. As such, undisinfected effluent samples contained relatively high ammonia-N concentrations (~10 mg/L as N, typically). Consequently, chlorination of Facility B effluent samples resulted in the formation of residual chlorine in the form of chloramines. Therefore, residual chlorine in chlorinated Facility B samples was comprised largely of NH_2Cl, with smaller quantities of $NHCl_2$ with traces of NCl_3 and organic chloramines. Inorganic chloramines are generally regarded as less effective antimicrobial agents than free chlorine (White, 1986). Organic chloramines have been shown to be generally ineffective disinfectants (Donnermair and Blatchley, 2003). Therefore, it is not surprising that the concentration of viruses in the chlorinated effluent samples, as well as the viral species diversity, were relatively high, particularly as compared with samples that had been subjected to UV irradiation. In general, UV irradiation was a more effective antiviral strategy than chlorination/dechlorination in the non-nitrified samples from Facility B.

The nucleic acid content of the bacteriophage surviving chlorine disinfection was predominantly DNA (10 out of 12 isolates). Only two RNA bacteriophages were isolated from samples after chlorine disinfection using an *E. coli* host (Table 6-2).

Table 6-2. Nucleic Acid Content of Post-disinfection Viable Phage in Facility B Samples.

Disinfectant Exposure Scenario	Host Strain	Classification	
		DNA	RNA
40 min contact time; 2.0 mg/L as Cl_2	F+ amp	6	0
40 min contact time; 2.0 mg/L as Cl_2	*E. coli*	4	2
20 mJ/cm^2 UV	F+ amp	0	0
20 mJ/cm^2 UV	*E. coli*	0	0

6.3.3 Facility C Samples

Only two samples were assayed from Facility C due to the low number of bacteriophage in the secondary effluent samples (Figures 6-24 and 6-25). Coliphage were detected in both samples, while F+ bacteriophage were detected only in one sample at a concentration of 5 pfu/mL. No bacteriophage were detected after a UV dose of 20 mJ/cm^2 in either of the samples. Two plaques isolated after 40 minutes chlorine disinfection did not regrow after isolation and attempted high titer production using an *E. coli* host. Facility C is the only facility sampled during this study that performs nitrification and denitification as a part of its treatment scheme. This may have contributed to the low concentration of viable phage isolated from this facility.

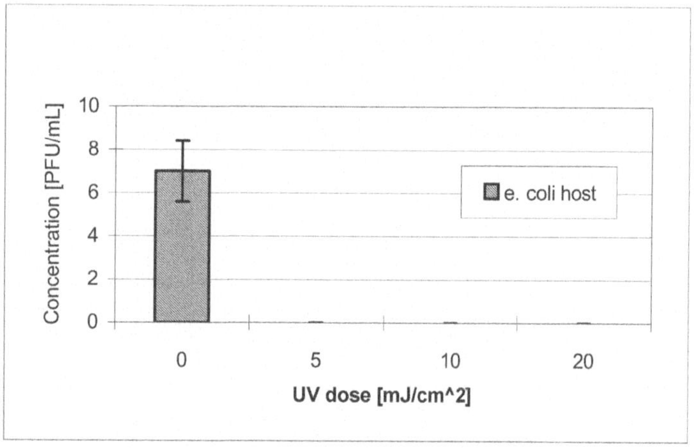

Figure 6-24. UV Inactivation of Indigenous Phage. Sample Collected from Facility C, October 17, 2001.

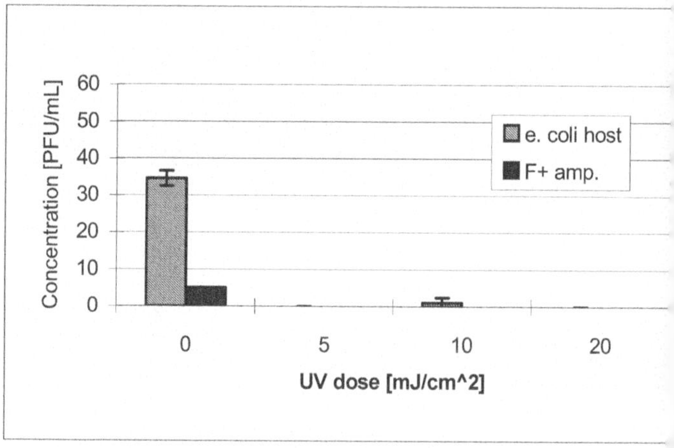

Figure 6-25. UV Inactivation of Indigenous Phage Using an *E. coli* Host. Sample Collected from Facility C, February 8, 2001.

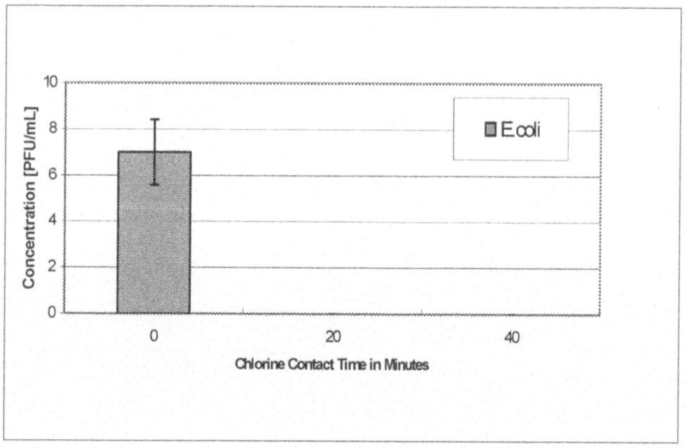

Figure 6-26. Chlorine Inactivation of Indigenous Phage Based on an Initial Chlorine Concentration of 2.0 mg/L (as Cl_2). Sample Collected from Facility C, October 17, 2001.

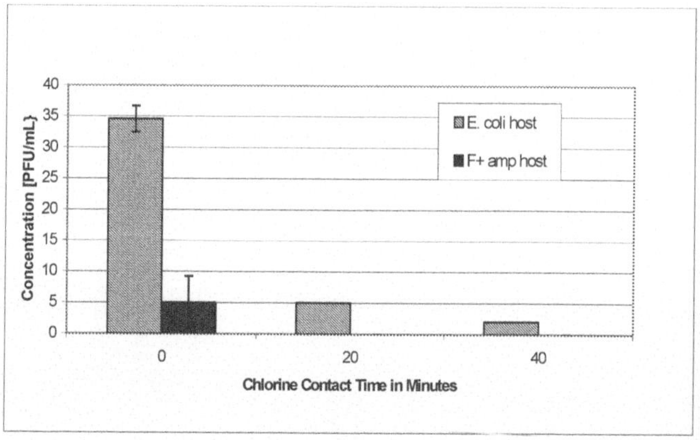

Figure 6-27. Chlorine Inactivation of Indigenous Phage Based on an Initial Chlorine Concentration of 2.0 mg/L (as Cl_2). Sample Collected from Facility C, July 8, 2001.

6.3.4 Facility D Samples

The indigenous bacteriophage concentrations in samples collected from Facility D were slightly higher than those detected at Facility C (Figures 6-28 through 6-34). These phage were made up predominantly of somatic bacteriophage (*E. coli* host) with very low concentrations of viable F+ bacteriophage. Filtration of samples using a 0.22 µm membrane did not appear to alter the UV sensitivity of the bacteriophage using either the *E. coli* or the F+ amp hosts (Figures 6-28 and 6-29).

Only one sample had any bacteriophage detected post exposure to a UV dose of 20 mJ/cm^2 (Figure 6-30). There were between 5 pfu/mL and 20 pfu/mL of somatic coliphage detected in samples post chlorine disinfection (Figures 6-33 through 6-35). Facility D employs suspended growth treatment without nitrification. As described previously, this situation dictates that residual chlorine will be in the form of chloramines, which are relatively ineffective antiviral agents.

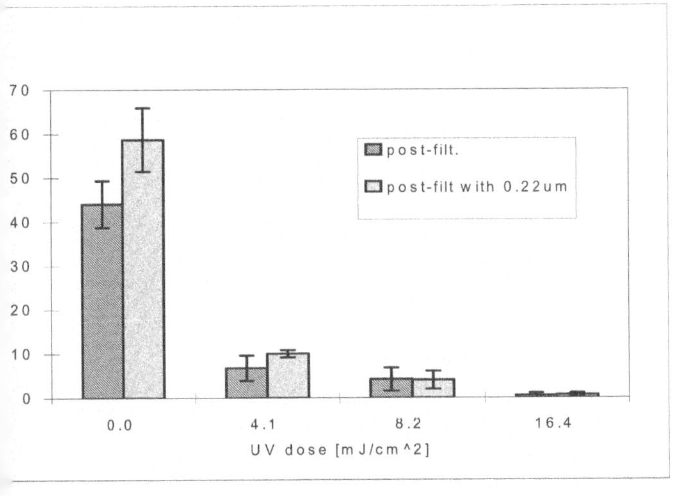

Figure 6-28. UV Inactivation of Indigenous Phage Using an *E. coli* Host. Sample Collected from Facility D, February 1, 2001.

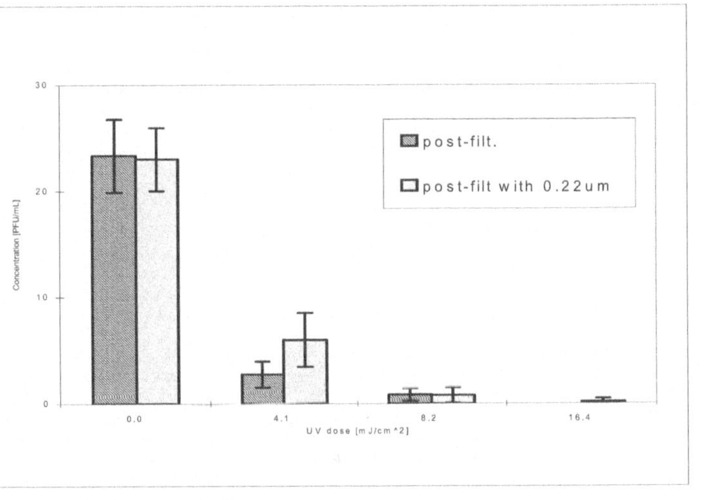

Figure 6-29. UV Inactivation of Indigenous Phage Using an F+ amp Host. Sample Collected from Facility D, February 1, 2001.

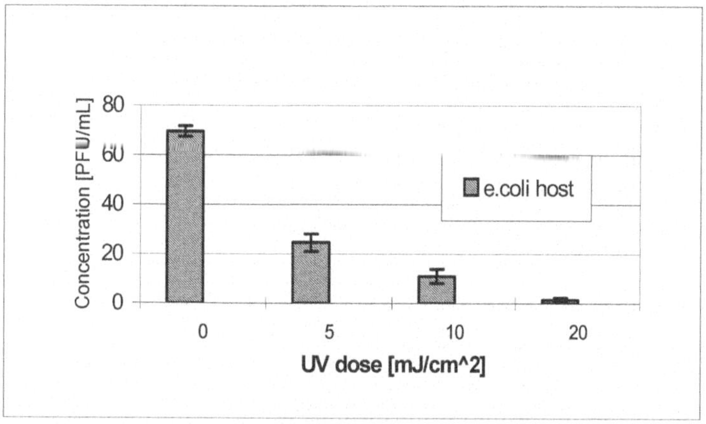

Figure 6-30. UV Inactivation of Indigenous Phage. Sample Collected from Facility D, April 3, 2001.

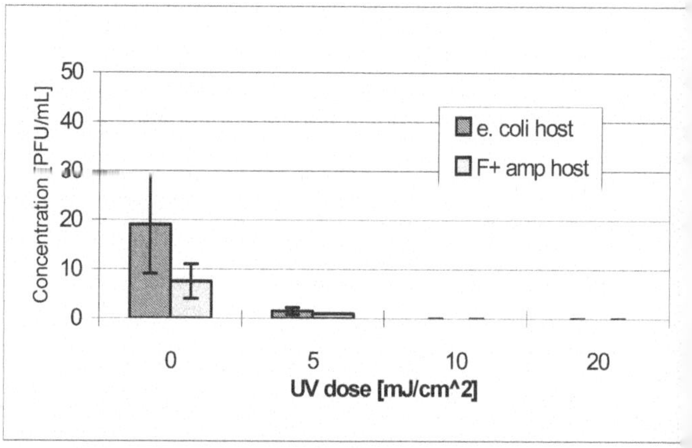

Figure 6-31. UV Inactivation of Indigenous Phage. Sample Collected from Facility D, August 21, 2001.

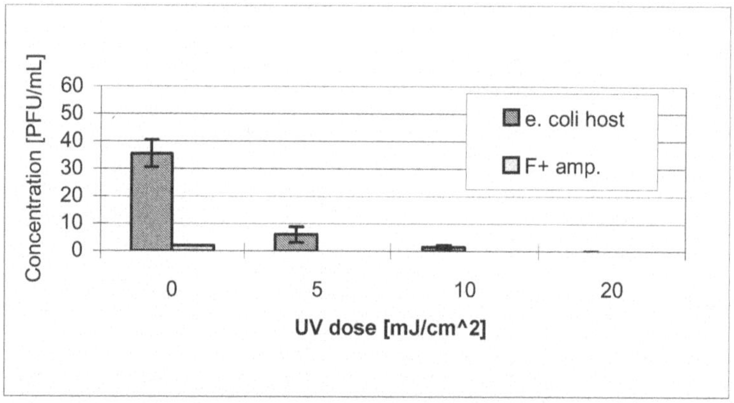

Figure 6-32. UV Inactivation of Indigenous Phage. Sample Collected from Facility D, September 6, 2001.

disinfection to be applied on a seasonal basis, based on the assumption that minimal direct human contact will occur during these cold-weather months. During warmer months, when virus concentrations may be increased, chlorine-based disinfection appears to be relatively ineffective as a method of reducing viable viral concentrations to assure the protection of public health from the impact of wastewater effluent discharge to receiving waters.

F-specific RNA bacteriophage belonging to the *Leviviridae* family have been shown to be among the most resistant organisms to wastewater disinfection, along with hepatitis A virus and rotavirus SA-11. The detection of F-specific RNA bacteriophage in chlorinated wastewater effluents could indicate the presence of human pathogenic viruses.

In contrast to chlorination/dechlorination, UV irradiation provides relatively consistent reduction of bacteriophage under a variety of wastewater conditions. Fluctuations in bacteriophage concentration due to mixtures of wastewater inputs did not appear to reduce the effectiveness of UV disinfection. Further, the consistent reduction of F-specific phage identified throughout this study provides insight into the reduction of enteric virus that could be achieved since F-specific bacteriophage are considered to be more resistant to UV inactivation than many enteric viruses.

CHAPTER 7.0

TOXICITY OF ALIPHATIC AMINES AND THEIR CHLORINATED FORMS

7.1 Background

The use of chlorine for wastewater disinfection has been identified as a contributor to effluent toxicity. Although S(IV)-based dechlorination practices are quite effective for reduction of free and inorganic combined chlorine residuals, some chlorinated compounds are resistant to dechlorination. Among these compounds, chlorinated organic amines are of particular concern.

The presence of organic amines in municipal wastewater effluent is largely determined by the sources of amines within the system, as well as the extent to which these compounds may be removed or chemically altered by treatment processes. Specific treatment operations that are most likely to influence the composition of organic-N within a municipal wastewater treatment system include nitrification and chlorination.

Experiments were conducted to characterize the chemistry and toxicological behavior of several model organic-N compounds. Since organic-N in municipal wastewater is likely to exist in the form of a mixture of many compounds, several simple aliphatic amines were chosen as model compounds to represent the compounds that could be present in municipal wastewater effluent samples.

7.2 Materials and Methods

7.2.1 Chemical Preparation

All solutions were prepared from reagent grade chemicals. Nanopure water was used in the preparation of all solutions and in sample dilution. Stock solutions of free chlorine were prepared from 4% NaOCl and diluted to 4.2 mM, then stored in aluminum foil-covered, glass-stoppered flasks. Concentrations of the stock solutions were determined titrimetrically before use.

Four model aliphatic amines were used for chlorination tests: methylamine, n-propylamine, ethylmethylamine, and trimethylamine. Figure 7-1 lists structures, as well as physical and chemical properties, of these model organic amines. Each amine was purchased (Sigma-Aldrich) and used without further purification. Aqueous solutions of the model organic-N compounds were stored in the dark at $4^{o}C$. Before chlorination, a stock solution of amine was prepared from the aqueous amine with a dilution ratio of 50. The stock solution was made before each experiment and was further diluted before chlorine addition using phosphate buffer to achieve the designed concentration and pH. After preparation, the stock solution was stored in a vial and refrigerated at $4^{o}C$ to reduce volatility.

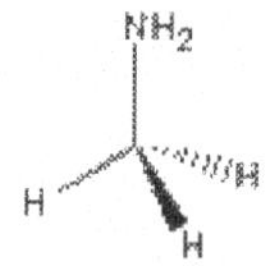

Methylamine (CH_5N, MW=31.0572)

Solubility[1]=1.08×10^6 mg/L at 25°C

Vapor pressure[1]=2650 mmHg at 25°C

pKa[1]=10.6

n-Propylamine (C_3H_9N, MW=59.1108)

Solubility[2]=1×10^6 mg/L at 25°C

Vapor pressure[2]=310 mmHg at 25°C

pKa[2]=10.7

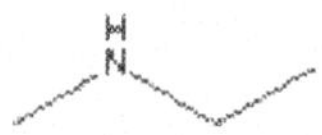

Methylethylamine (C_3H_9N, MW=59.1108)

Solubility[3]=1×10^6 mg/L at 25°C

Vapor pressure[3]=390 mmHg at 25°C

pKa[3]=N/A

Trimethylamine (C_3H_9N, MW = 59.1108)

Solubility[4]=8.9×10^5 mg/L at 25°C

Vapor pressure[4]=1607 mmHg at 25°C

pKa[4]=9.8

1. Hhttp://esc.syrres.com/interkow/webprop.exe?CAS=74-89-5H
2. Hhttp://esc.syrres.com/interkow/webprop.exe?CAS=107-10-8H
3. Hhttp://esc.syrres.com/interkow/webprop.exe?CAS=624-78-2H
4. Hhttp://esc.syrres.com/interkow/webprop.exe?CAS=75-50-3H

Figure 7-1. Structure, Physical, and Chemical Properties of Model Organic-N Compounds Used in this Research.

7.2.2 Model Organic Chloramines

Standard organic chloramine solutions were prepared by addition of free chlorine to precursor stock solutions of the above model amine solutions. Each precursor solution was added into an aqueous solution of free chlorine at a chlorine:precursor (Cl:P) molar ratio of 0.925:1. The initial concentration of each precursor in the sample was 10 mg/L. Each organic chloramine solution was mixed in an aluminum foil-covered, glass-stoppered flask for 15 minutes to promote equilibration. Ferriol et al. (1991) indicated that chlorine transfer from monochloramine to aliphatic amines was fast and the rate constants varied with pH values, with the maximum values occurring at pH 9.0 to 10.5 for the amines tested. To facilitate chlorine transfer, pH conditions chosen in this research were similar to the study of Ferriol et al. (1991). Organic chloramine solutions were used within two hours of preparation.

DPD titration was used to determine the concentration of the resulting organic chloramine and free chlorine to ensure that no free available chlorine remained in solution after reaction. MIMS was used to identify the chemical structures of resulting organic chloramines and ensure that precursors reacted completely with free chlorine. Detailed descriptions regarding the setup and configuration of the MIMS system can be found in Shang and Blatchley (1999) and Shang et al. (2000). The system was operated under electron impact ionization of 70 eV using scan mode (m/z=41–200) to evaluate formation of organic chloramines. Each precursor was found to react rapidly with free chlorine, and the resulting organic chloramine was stable and abundantly present in solution.

7.2.3 Chlorination of Aliphatic Amines

7.2.3.1 DPD Titration

DPD titration is one of the most common analytical methods used for measurement of +1-valent chlorine compounds. It has also been adopted as a standard method for residual chlorine and chloramine measurement in drinking water and wastewater (*Standard Methods*, 1995).

The chemical basis of this technique is oxidation of DPD by chlorine to yield two oxidation products (Harp, 2002). At near neutral pH, chlorine reacts with DPD to produce a semiquinoid cationic compound known as Würster dye, which exhibits the magenta color that is characteristic of the DPD colorimetric test. DPD can be further oxidized to an imine compound, which is relatively unstable and colorless. When present at low concentrations at near-neutral pH, chlorine reacts with DPD to yield the Würster dye as the principal oxidation product. At higher oxidant levels, the formation of the unstable colorless imine is favored, resulting in fading of the colored solution.

The presence of inorganic chloramines interferes with the determination of free chlorine in DPD titration since DPD also reacts directly with free chlorine and monochloramine (Cooper et al., 1975; Guter et al., 1974). Furthermore, free chlorine can react with organic-N compounds to form a variety of organic chloramines that will cause false positive bias for the determination of free chlorine (Ram and Morris, 1980; Ram and Malley, 1984; Jensen and Johnson, 1990a,b). Conventional methods for measuring chloramines cannot distinguish between inorganic chloramines and some organic chloramines (Isaac et al., 1980; Wolfe et al., 1985; Yoon and Jensen, 1995). Excess sulfite used for dechlorination reduces the iodine formed in the titration reaction directly, providing an incorrectly low value of residual chlorine (Nweke and Helz, 1990).

To overcome the problems associated with DPD titration, alternative analytical methods have been developed and published. These methods include derivatization/HPLC (Lukasewycz et al., 1989), membrane separation/spectrophotometric measurement (Aoki, 1989), and a stopped-flow spectrophotometric method (Schurter et al., 1995). These methods provide reproducible quantitative data, but still suffer from drawbacks and limitations. Some of these methods involve sample preparation steps that are time consuming and can change equilibrium among chlorine species. Some of the methods are not suitable for application in samples composed of complex mixtures (such as wastewater).

Due to the limitations of DPD titration described above, MIMS was used for characterization of organic chloramines. Composition and possible structural information of organic chloramines can also be obtained by MIMS.

7.2.3.2 MIMS

MIMS is one of the most efficient, simple, and sensitive techniques for the analysis of volatile compounds in water (Wong et al., 1995; Kotiaho et al., 1991a,b). The MIMS method has been demonstrated to be superior to conventional titrimetric methods (e.g., DPD/FAS or amperometric) for monitoring free chlorine and inorganic chloramines during chlorination because it is less subject to analytical interference and has similar limits of detection.

In MIMS, a membrane is used as a separation device. Analytes diffuse through the membrane and pass into a moving gas phase, which in turn will flow into the ionization source of the mass spectrometer. Sample introduction can be continuous or intermittent. Therefore, MIMS is capable of online, nearly real-time monitoring. The capability of providing nearly real-time information represents a significant advantage over conventional techniques for chlorine quantification. Further, samples can be directly delivered to the membrane interface; generally no pre-concentration or other workup prior to analysis is required. MIMS is characterized by other important analytical advantages relative to other methods, such as low detection limits (Bauer and Solyom, 1994) and relatively short response times. By proper modification with combination of fast gas chromatography, MIMS systems can achieve GC separation while maintaining the online monitoring characteristics (Chang and Her, 2000).

The MIMS method allows essentially unambiguous quantification of inorganic combined chlorine (i.e., it is not subject to interference by organic chloramines). Though it is likely that the various forms of residual chlorine measured by MIMS will display different antimicrobial efficacy, it is not possible at present to separate or quantify the effects of these constituents. In this research, MIMS was applied mainly for detection and identification of volatile DBPs. Identification of DBPs of the four model compounds after chlorination was achieved using scan mode (m/z=41-200) on the MIMS system.

The MIMS system was modified from a Hewlett-Packard (HP) 5892 benchtop GC/MS, containing an HP 5972A single quadrupole mass selective detector (MSD) and a computer control program (HP G1034C Software as the MS ChemStation). A flow-through membrane cell was installed in the oven for precise and quick temperature control. Analytes were introduced and delivered to the membrane cell by a peristaltic pump and a Rheodyne 9010 injector was used to reduce flow disturbances. To reduce the likelihood of contamination, the pump was located downstream of the flow system. The membrane cell was constructed by connecting a ¼-in. o.d., 5-cm long glass tube with two T unions (Figure 7-2).

Samples were pumped through the inside of the membrane since the ratio of surface area to the volume of the solution can be enhanced when sample solution flowing inside of the membrane, thereby increasing the rate of analyte transport through the membrane. A 7-cm long, 0.25 mm i.d., 0.47 mm o.d., silicon hollow membrane (Baxter, Medical/Pharmaceutical Grade) was connected to 1/16-in. o.d. tubing and mounted in the center of the membrane cell. The exposed length of the hollow silicone membrane in the membrane cell was approximately 6 cm. The characteristics of this tubing and the principles of analyte transport and separation are described by Kotiaho et al. (1991 a,b). A flow restrictor made from 0.1 mm i.d., 1-m long tubing was connected at the branch outlet of one T union to deliver helium gas to the membrane cell. The liquid/gas flow was operated in a counter-current manner to achieve maximum mass transfer. Absolute pressure in the ion source chamber was maintained at 10^{-5} Torr. The 1/16-in. o.d. tubing, connected at the branch of the other T union, was used to transfer permeable analytes with carrier gas into the ionization source. All tubing, ferrules, and joints that come into contact with analytes were made of Teflon PFA or stainless steel to limit reactions and possible corrosion. System control, tuning, data acquisition, and data analysis were managed with MS ChemStation software.

The system was operated under electron impact ionization of 70 eV. Mass spectra scan mode was used to evaluate formation of organic chloramines and other possible intermediates while selected ion monitoring (SIM) mode was used for quantification of specific compounds.

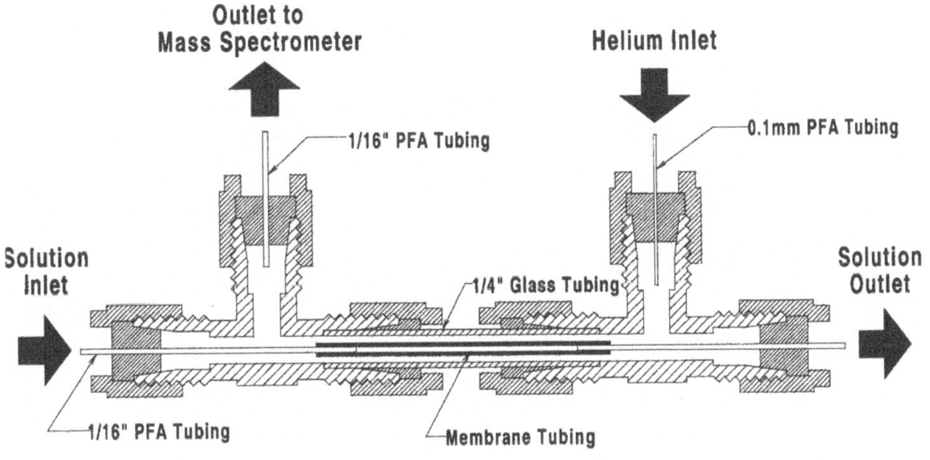

Figure 7-2. Schematic Representation of Membrane Interface Used in MIMS System.

7.2.3.3 Experimental Procedures

High concentrations of amines (10 or 100 mg/L as N for each batch reaction) were used in MIMS experiments to facilitate product detection, and increase reaction rates. Volatile DBP formation was monitored in real-time by MIMS during the first 10 minutes of chlorination, and intermittently at subsequent reaction times to test product stability and formation of other DBPs with slow reaction rates.

Prior to sample injection, the MIMS system was stabilized by pumping nanopure water through the membrane cell to develop a steady background signal. After stabilization with nanopure water, the online reaction monitoring experiment was initiated first by introduction of a free chlorine solution containing pH buffer, then the addition of model organic amine solution. The membrane cell of the MIMS system was maintained isothermally at 70°C and temperature of the ion source was controlled independently at 190°C.

7.2.4 Toxicity of Model Aliphatic Amines and Their Chlorinated Forms

7.2.4.1 Microtox

A Microtox Model 500 analyzer was used to assess the acute toxicity of organic chloramines. Tests were performed as described in the Acute Detailed Basic Testing Procedure (Microbic Corporation, 1992; Ribo and Kaiser, 1987; Kaiser and Kaiser, 1988).

A marine bacterium *Phosphobacterium phosphoreum* (also known as *Vibrio fischerii*, strain no. NRRL-B-11177) was used for the Microtox acute toxicity assay. The bioluminescent organisms were rehydrated with reconstitution solution from a lyophilized freeze-dried bacterial culture and the rehydrated culture was then incubated at 5.5±1°C for 20 minutes to achieve a stable condition. The acute toxicity assay involved exposing the stabilized bioluminescent bacteria to test samples under different concentrations, followed by measurement of light output of the luminescent bacteria after a given exposure time. A decrease in bioluminescence was assumed to correlate with acute toxicity. All samples were incubated at 15±0.5°C during exposure.

The Microtox acute toxicity assay has been applied for monitoring of effluents and sediment contamination to comply with environmental protection regulations in several countries because it is a relatively fast, low-cost analysis (D.EPS, 1990). Further, the Microtox test, where sufficient data are available, generally shows positive correlations between the effects on *P. phosphoreum* and other toxicity bioassays, including: rainbow trout (*Oncorhynchus mykiss*), fathead minnow (*Pimephales promelas*), and waterfleas (*Daphnia magna*) (Blum and Speece, 1991; Nacci et al., 1986).

Test samples used in this study included precursor aliphatic amines (n-propylamine, ethylmethylamine, and trimethylamine) and chlorinated amines (n-chloropropylamine, chloroethylmethylamine, and chlorodimethylamine). The concentration of each chlorinated amine was quantified by DPD titration, and the structure of each chlorinated amine was validated by MIMS. Because of the high solubility of these amines, samples were diluted serially with nanopure water. Toxicity test results are presented in the form of EC_{50} values with 5 or 15-minute exposure. Three or more toxicity test replicates were performed for each sample.

7.2.4.2 Oxygen Utilization Rate Test

The oxygen utilization rate (OUR) test was carried out on a Gilson Oxygraph respirometer (Model 5/6H) with the use of an enhanced nitrifier culture. The nitrifiers were fed, cultured, and incubated in an aluminum foil-covered, 18-L bench-scale flow-through chemostat, at 25°C. The culture was maintained by the addition of ammonium chloride as a substrate and sodium bicarbonate for pH adjustment. Suspension pH was controlled between 7.1 and 7.3. The resulting mixed liquor suspended solid (MLSS) concentration was approximately 2900 mg/L. Prior to each OUR assay, an aliquot of the nitrifier suspension was transferred into a glass flask with aluminum foil and stabilized under aeration and mixing overnight to consume the residual ammonium-related substrate. After stabilization of the nitrifier, amine or organic chloramine was introduced along with the nitrifiers into a 2-mL opaque respiration cell. Each sample was exposed for 3 minutes under constant mixing strength prior to oxygen uptake analysis. After three minutes, 50 μL of substrate made of ammonium chloride (1000 mg/L as N) was injected into the respiration cell and oxygen concentration in the sample was recorded with time. Both chlorinated and unchlorinated amines were subjected to the OUR test to quantify respiratory inhibition. Blank samples were assayed before each test to ensure the stability of the nitrifier suspension.

7.3 Results and Discussion

7.3.1 Chlorination of Aliphatic Amines

7.3.1.1 Methylamine

For each batch reaction, pH was controlled by 0.01 mM borate buffer (pH, 9.20), phosphate buffer (pH, 6.0), or by addition of 1 mL of 10 N NaOH in 100 mL of sample to achieve a pH value >13. Table 7-1 lists the molar ratios of model organic-N compounds/chlorine used. Table 7-2 lists the summary of major characteristic peaks for chlorination of each model aliphatic amine. The mass spectra presented in this part were obtained using scan mode (m/z=41–200) with electron ionization. (EI).

Chlorination of methylamine was examined at chlorine:nitrogen ratios (Cl:N) of 0.5 and 10. MIMS was used to analyze solution composition as a function of time. Figures 7-3 and 7-4, respectively, show representative mass spectra of chlorinated methylamine under low and high Cl:N ratios. For the low Cl:N ratio (0.5), the presence of peaks of m/z=65 and m/z=67 with a ratio of 3:1 is consistent with chlorine substitution for hydrogen. A possible structure corresponding to these peaks is monochloromethylamine. For the high Cl:N ratio (10), the presence of peaks of m/z=99, m/z=101, and m/z=103, with a ratio of 9:6:1, was consistent with a di-chlorine-substituted product. A possible structure corresponding to these peaks is dichloromethylamine. No evidence of chlorine substitution for the methyl group was observed in the chlorination of methylamine, even after a long reaction time (Figure 7-4). Based on these results, the products of methylamine chlorination were reasonably stable under these conditions, and the maximum number of chlorine substitutions in methylamine was two. Chlorine substitution in this system appears to be limited to the amino-N hydrogens. Under high Cl:N ratio, a significant peak with m/z=62 was observed. Since there was no fragment ion of dichlormethylamine or monochloromethylamine corresponding to this peak (Table 7-2), and the relative abundance at m/z=62 increased with reaction time, this may suggest the formation of CH_2NCl, a product resulting from loss of HCl from dichloromethylamine.

Table 7-1. Summary of Chlorination Conditions for Methylamine.

Sample No.	Cl:N Molar Ratio	PH	Buffer Type
A1[1]	0.5:1	9.20	Borate, 0.02M
B1[2]	10:1	9.20	Borate, 0.02M
A2[1]	0.5:1	9.20	Borate, 0.02M
B2[2]	10:1	9.20	Borate, 0.02M
C1	10:1	> 13	N/A[4]
D1[3]	10:1	6.0	Phosphate, 0.02M
D2[3]	10:1	6.0	Phosphate, 0.02M

*1. A1, A2 use the same sample but with different reaction time.
2. B1, B2 use the same sample but with different reaction time.
3. D1, D2 use the same sample but with different reaction time.
4. No buffer was added, but 2 mL 10N NaOH was added to 100 mL of sample.

Table 7-2. Summary of Product Ion Spectra of Different Forms of Chlorinated Aliphatic Amines.

Precursor	No. of Cl substitution	Chlorinated form	Parent ion m/z	Ions characteristic of chlorination position (m/z)					
				M-H*	M-Cl	M-HCl	$M-CH_3$	$M-CH_3-Cl$	$M-C_2H$
H_2NCH_3	1	$H^{35}ClNCH_3$	65	64					
	1	$H^{37}ClNCH_3$	67	66					
	2	$^{35}Cl_2NCH_3$	99	98	64	63	84	49	
	2	$^{35}Cl^{37}ClNCH_3$	101	100	66	65	86	51	
					64	63		49	
	2	$^{37}Cl_2NCH_3$	103	102	66	65	88	51	
$H_2NC_3H_7$	1	$H^{35}ClNC_3H_7$	93	92					64
	1	$H^{37}ClNC_3H_7$	95	94					66
	2	$^{35}Cl_2NC_3H_7$	127						98
	2	$^{35}Cl^{37}ClNC_3H_7$	129						100
	2	$^{37}Cl_2NC_3H_7$	131						102
$CH_3NHC_2H_5$	1	$CH_3N^{35}ClC_2H_5$	93	92			78	43	64
	1	$CH_3N^{37}ClC_2H_5$	95	94			80	43	66
$N(CH_3)_3$	1	$^{35}ClN(CH_3)_2$	79	78					
	1	$^{37}ClN(CH_3)_2$	81	80					
	2	$^{35}Cl_2NCH_3$	99	98	64	63	84	49	
	2	$^{35}Cl^{37}ClNCH_3$	101	100	66	65	86	51	
					64	63		49	
	2	$^{37}Cl_2NCH_3$	103	102	66	65	88	51	

* M: molecular ion

After chlorination, the sample was stored in a stoppered glass flask in the dark for 24 hours under high and low Cl:N ratios to test product stability. Results are summarized in Figures 7-3b and 7-4b. From these spectra, there was no new peak observed after long reaction time and the abundances of chlorinated products after long reaction time were close to those with short reaction time (10 minutes), suggesting that the reaction reached completion quickly and the resulting products were reasonably stable, at least on a 24-hour time-scale. After a reaction time of 22 days (Figure 7-4c), the abundance signal corresponding to dichloromethylamine (m/z=98) decreased to 46.7% of its value at t=10 minutes.

Different pH values (6.0, 9.2, >13) were used for chlorination of methylamine under high Cl:N ratio (10), and the results are summarized in Figures 7-6a, 7-4a, and 7-5, respectively. Since the pK_a of HOCl is 7.5, the pH values of 6 and 9.2 were chosen to test the difference of different chlorination reagents (HOCl, OCl^-) on chlorination results. Also, since the pK_a for methylamine is 10.6, most of the methylamine would present in protonated form under pH 6.0 and 9.2. The pH

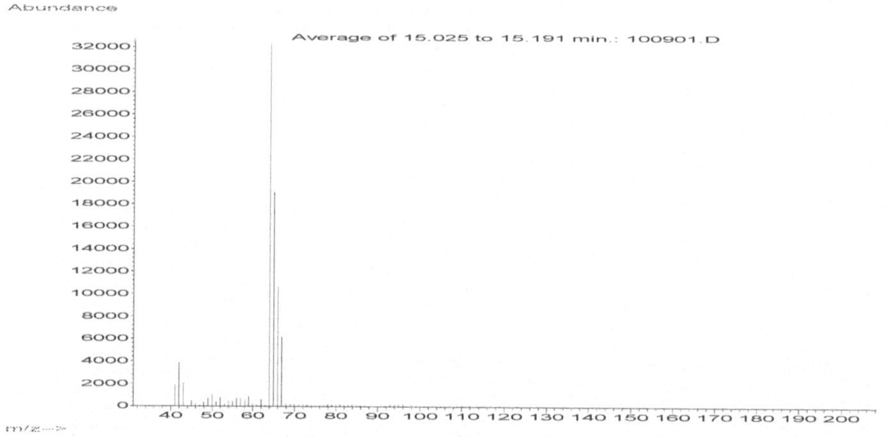

(a) Spectrum A1 Cl:N = 0.5:1, pH = 9.20, reaction time = 10 min

Figure 7-3a. Chlorination of Methylamine Under Low Cl:N Ratio (0.5) and pH = 9.20; Reaction Time=10 Minutes.

value of >13 was chosen to test the chlorination behavior of neutral and protonated methylamine, though this pH is unlikely to be encountered in wastewater chlorination. The resulting spectra showed that there was no significant difference between pH=6 and pH=9.2 with reaction time of 10 minutes, suggesting the chlorination ability of hypochlorous acid and the hypochlorite ion are similar toward methylamine. Under high pH conditions (>13), a substantial difference was observed at peaks with m/z = 64 to 67. By comparing with Table 7-2, monochloromethylamine can also contribute to these ions; the increase of abundance in these ions may suggest that formation of monochloromethylamine is more favorable at high pH. From the kinetic viewpoint, high pH values will favor the presence of organic chloramines in their neutral forms and increase the rate of the HCl loss from dichloromethylamine, resulting in higher yield of CH_2NCl (Pedersen et al., 1999). The significant increase in abundance of m/z=62 from pH 9.20 to pH >13 may reflect a different yield of CH_2NCl. A similar increase in abundance at peaks with m/z 62 and 64 was observed under pH=6.0 with a long reaction time (22 days) (Figure 7-6b). From Figure 7-6, it is clear that further reaction took place since a change of relative abundance at m/z=62 and 64 was observed.

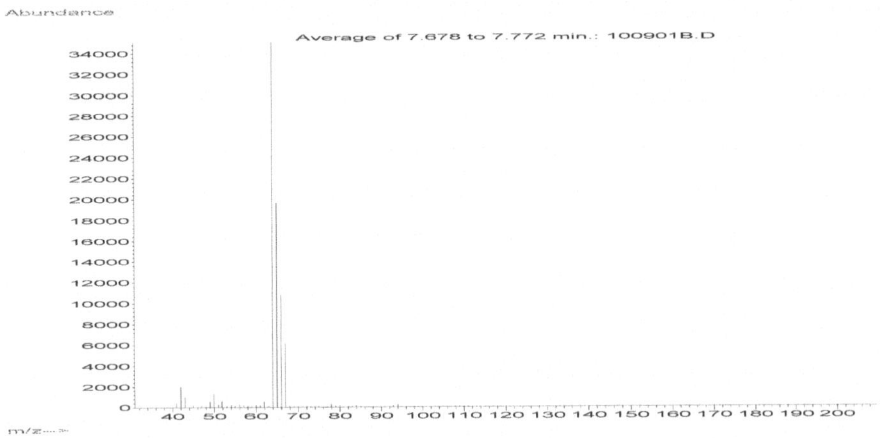

(b) Spectrum A-2 Cl:N = 0.5:1, pH = 9.20, reaction time = 24 hr

Figure 7-3b. Chlorination of Methylamine Under Low Cl/N Ratio (0.5) and pH = 9.20; Reaction Time=24 Hours.

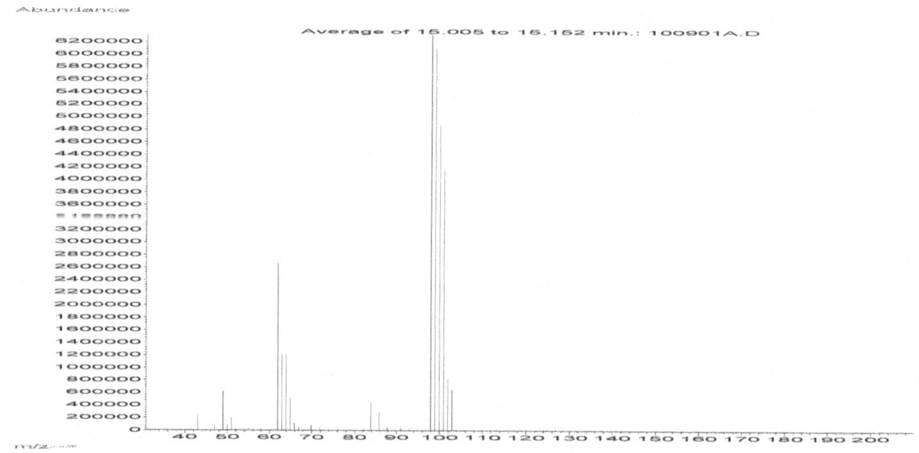

(a) Spectrum B1 Cl:N = 10:1, pH = 9.20, reaction time = 10 min

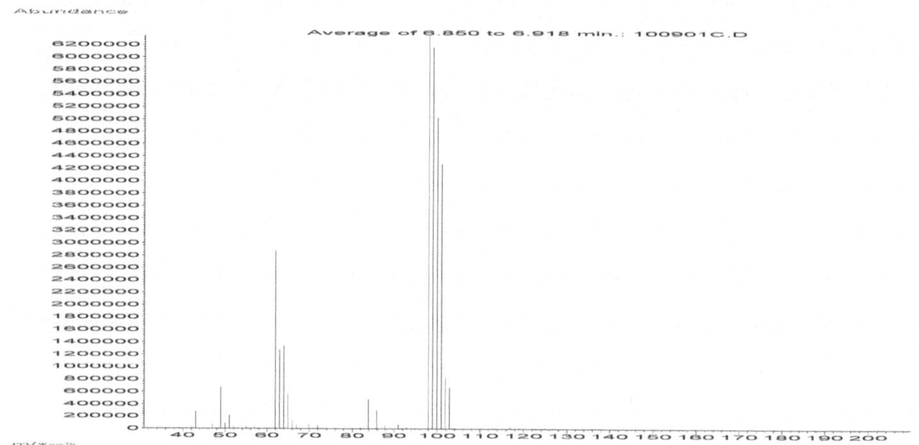

(b)Spectrum B2 Cl:N = 10:1, pH = 9.20, reaction time = 24 hr

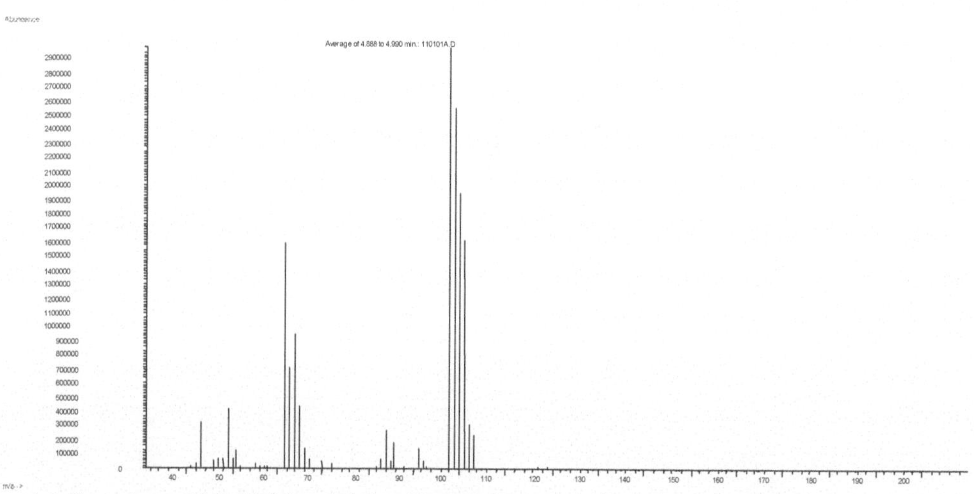

(c) Spectrum B3 Cl:N = 10:1, pH = 9.20, reaction time = 22 days

Figure 7-4. Chlorination of Methylamine Under High Cl:N Ratio (10) and pH = 9.20. (a) Reaction Time=10 Minutes; (b) Reaction Time=24 Hours; (c) Reaction Time=22 Days.

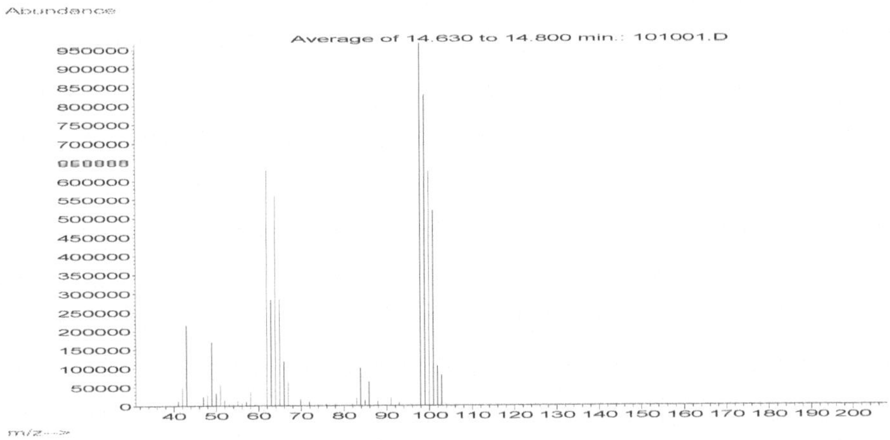

Spectrum C1 Cl:N =10:1, pH > 13, reaction time = 10 min

Figure 7-5. Chlorination of Methylamine Under High Cl:N Ratio (10) and High pH (>13).

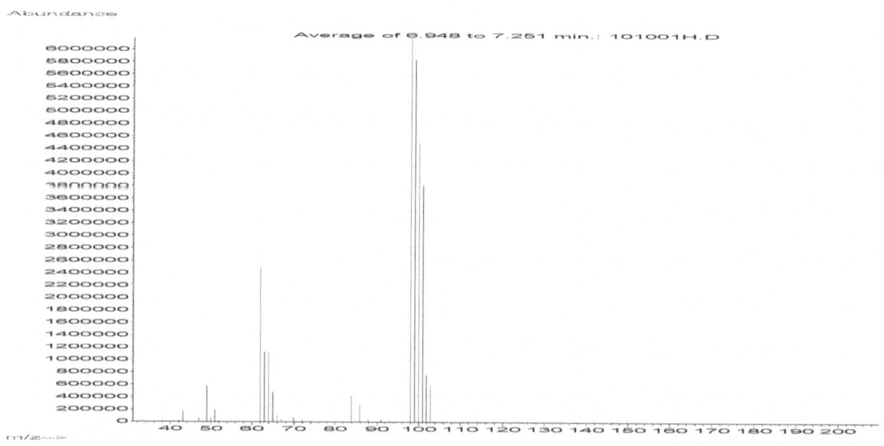

(a) Spectrum D1 Cl:N =10:1, pH = 6.0, reaction time = 10 min

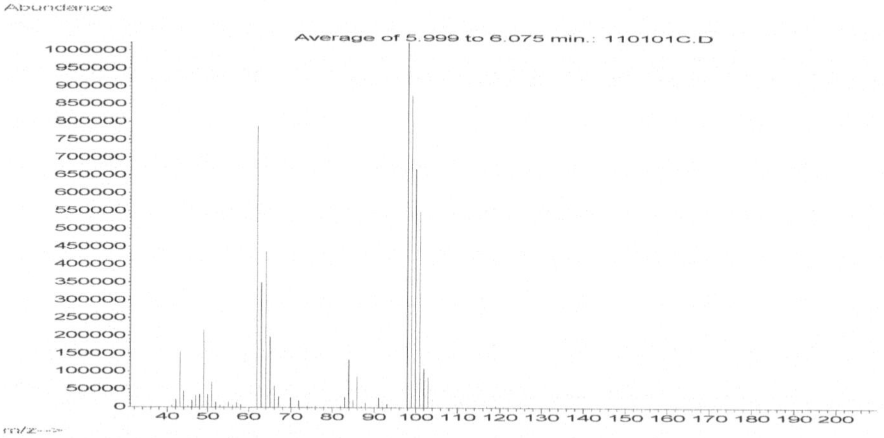

(b) Spectrum D2 Cl:N =10:1, pH = 6.0, reaction time = 22 days

Figure 7-6. Chlorination of Methylamine Under High Cl/N Ratio (10) and pH = 6.0. (a) Reaction Time=10 Minutes; (b) Reaction Time=22 Days.

7.3.1.2 Three Aliphatic Amine Isomers - n-propylamine, Ethylmethylamine and Trimethylamine

For each batch reaction, pH was either controlled by 0.01 mM borate buffer (pH, 9.20) or without addition of pH buffer. Table 7-3 lists the molar ratios of model organic-N compounds/ chlorine used. The mass spectra presented in this part were obtained using scan mode (m/z=41–200) with EI.

Figure 7-7 shows the mass spectra for chlorination of n-propylamine. For chlorination of n-propylamine under Cl:N ratio of two, the resulting spectrum, with isotope peaks of m/z=127, m/z=129, and m/z=131 with a ratio of 9:6:1, suggest two chlorine atoms were included in the chlorinated product. A reasonable compound with corresponding peaks is dichloropropylamine. Another cluster of peaks at m/z=98, m/z=100, and m/z=102 with a ratio of 9:6:1 suggests the loss of an ethyl group from dichloropropylamine. Table 7-2 lists other major peaks of fragment ions. The peaks at m/z=64 and m/z=66 under Cl:N=2 may come from monochloropropylamine, or possibly from monochloroethane. Under Cl:N ratio of 0.925, the resulting spectrum with isotope peaks at m/z=64, m/z=66 and m/z=93, m/z=95 with a ratio of 3:1 suggests one chlorine was included in the resulting product. A reasonable representation of the resulting product would be monochloropropylamine, which would have isotope peaks of molecule ion at m/z=93 and m/z=95 and fragment ions at m/z=64 and m/z=66 due to cleavage of C_2H_5 at the α position.

Figure 7-8 shows the mass spectra for chlorination of ethylmethylamine. At a Cl:N ratio of 0.925, the resulting spectrum with isotope peaks at m/z=78, m/z=80, and m/z=93, m/z=95 with a ratio of 3:1 suggests at least one chlorine was included in the resulting product. A reasonable guess of the resulting product would be monochloroethylmethylamine, which would yield isotope peaks of molecule ion at m/z=93 and m/z=95 and fragment ions at m/z=78 and m/z=80 due to cleavage of CH_3 at the α position. From those spectra, it is clear that change of Cl:N ratio from 0.925 to 10 did not produce any new peak; the maximum number for chlorine substitution in ethylmethylamine is 1.

Figure 7-9 shows the mass spectra for chlorination of trimethylamine. At a Cl:N ratio of 0.925, the resulting spectrum with isotope peaks at m/z=79, m/z=81 with a ratio of 3:1 suggests one chlorine was included in the resulting product. A possible explanation of the resulting product would be monochlorodimethylamine, which would yield isotope peaks of molecular ion at m/z=79 and m/z=81 due to the substitution of a methyl group by chlorine. At a higher Cl:N ratio (two), a group of isotope peaks showed up at m/z=99, m/z=101, and m/z=103, with a ratio of 9:6:1, suggesting di-chlorine substitution. Judging from the ratio of isotope peaks and its associated fragment ions, it appears that dichloromethylamine may be a possible product. It should be noted that m/z=78, m/z=79, m/z=80, m/z=81 may be attributable to chlorodimethylamine. It is observed that even at a high Cl:N ratio of 10, strong abundances were observed in peaks with m/z=78, m/z=79, m/z=80, and m/z=81, which correspond to the characteristic ions of chlorodimethylamine. This suggests the substitution for the second methyl group by chlorine is difficult, even in the presence of excessive free chlorine. However, no sign of substitution for the third methyl group by chlorine was observed (peak at m/z=117 was observed, however, the ratio of isotope peaks at m/z=117, m/z=119, m/z=121, and m/z=123 suggest it is not NCl_3), even when the Cl:N ratio was increased to 10. The maximum number of Cl substitutions in trimethylamine appears to be two, and the substitution may occur for a methyl group instead of hydrogen. It is interesting to note that under a Cl/N ratio > 2, a group of peaks was observed at m/z=117, m/z=119, m/z=121, and 123, and their abundance increased with reaction time. From the abundance ratio of these peaks, the data suggest the formation a product with at least three chlorine atoms. These peaks were similar to the characteristic peaks of carbon tetrachloride, suggesting a possible formation of carbon tetrachloride from chlorination of trimethylamine.

Recent studies (Mitch and Sedlak, 2002; Choi and Valentine, 2002) have characterized formation of N-nitrosodimethylamine (NDMA) from reaction of dimethylamine and monochloramine or sodium hypopchlorite. NDMA is a probable human carcinogen with a theoretical 10^{-6} lifetime risk level of cancer of 0.7 ng/L (U.S. EPA, 1997). In their proposed reaction scheme, dimethylchloroamine (DMCA) was formed as a precursor for NDMA. Chlorination of other secondary amines may also lead to NDMA formation. Dimethylchloroamine may also be formed by reaction of hypochlorite with trimethylamine (Section 7.4.2), though no evidence of NDMA formation was observed at a contact time of 300 minutes (Figure 7-9b). This may be due to slow reaction. However, the MIMS configuration used in this research does not have high sensitivity for NDMA (unpublished data).

In wastewater chlorination, pH is typically maintained in the range of 6 to 8 and contact times are usually <60 minutes. Under this pH range, aliphatic amines are likely to be present in their protonated forms. Also, judging from previously defined kinetics (Pedersen et al., 1999), loss of HCl from the chlorinated product would not be favored under this pH range. This suggests that organic chloramines will be relatively stable under the pH conditions that are typical in wastewater chlorination. Further toxicity investigations of these aliphatic amines were conducted, and the results are described in Section 7.5.

Table 7-3. Summary of Chlorination Conditions for N-propylamine, Ethylmethylamine, and Trimethylamine.

Sample	Sample No.	Cl:N Molar Ratio	PH	Buffer Type
N-propylamine	1-3	2	9.20	Borate, 0.02M
	1-3a[1]	2	9.20	Borate, 0.02M
	1-4	0.925	-	no buffer
Ethylmethylamine	2-3	2	9.20	Borate, 0.02M
	2-4	0.925	-	no buffer
	2-5	10	-	no buffer
Trimethylamine	3-3	2	9.20	Borate, 0.02M
	3-3a[2]	2	9.20	Borate, 0.02M
	3-4	0.925	-	no buffer
	3-5	10	-	no buffer

*1. 1-3, 1-3a use the same sample but with different reaction time.
2. 3-3, 3-3a use the same sample but with different reaction time.

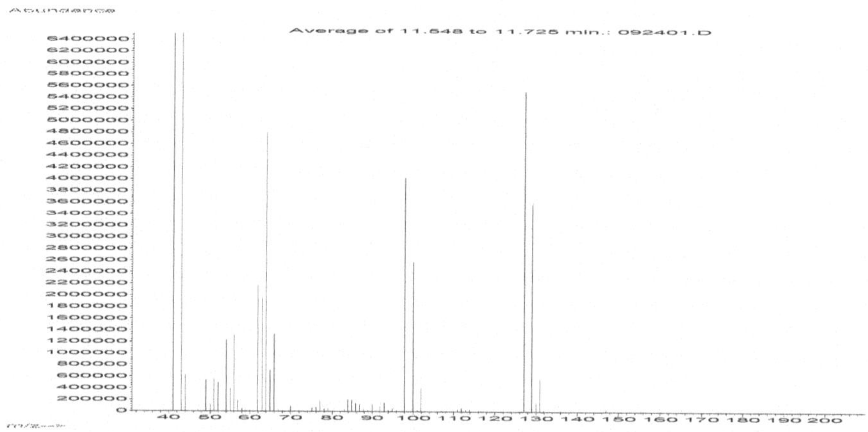

(a) Spectrum 1-3 Cl:N =2:1, pH = 9.20, reaction time = 5 min

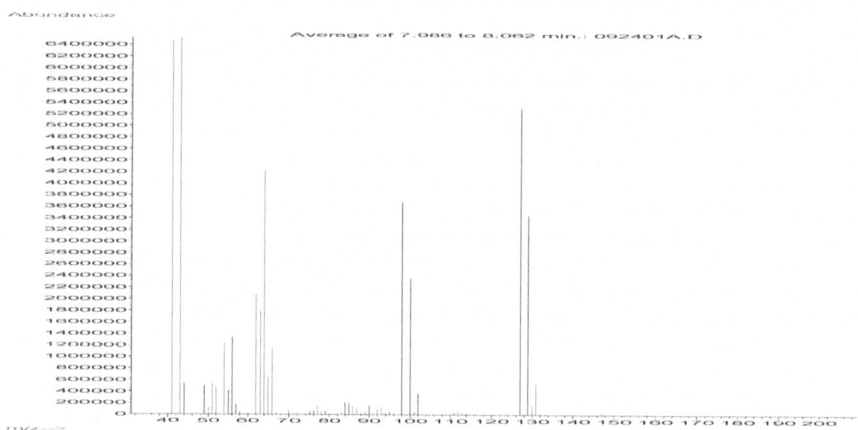

(b) Spectrum 1-3a Cl:N =2:1, pH = 9.20, reaction time = 300 min

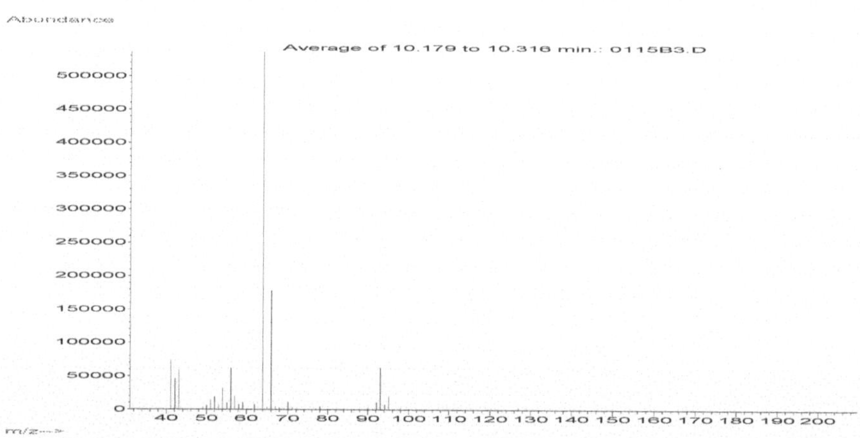

(c) Spectrum 1-4 Cl:N =0.925:1, w/o buffer, reaction time = 6 min

Figure 7-7. Chlorination of N-propylamine at pH=9.20 and Different Cl:N Ratio. (a) Cl:N=2, Reaction Time=5 Minutes; (b) Cl:N=2; Reaction Time=300 Minutes; (c) Cl:N=0.925, Reaction Time=6 Minutes.

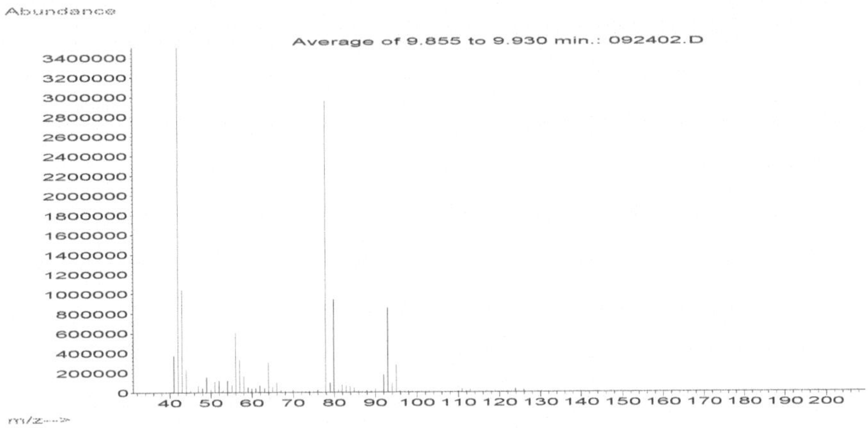

(a) Spectrum 2-3 Cl:N =2:1, pH = 9.20, reaction time = 6 min

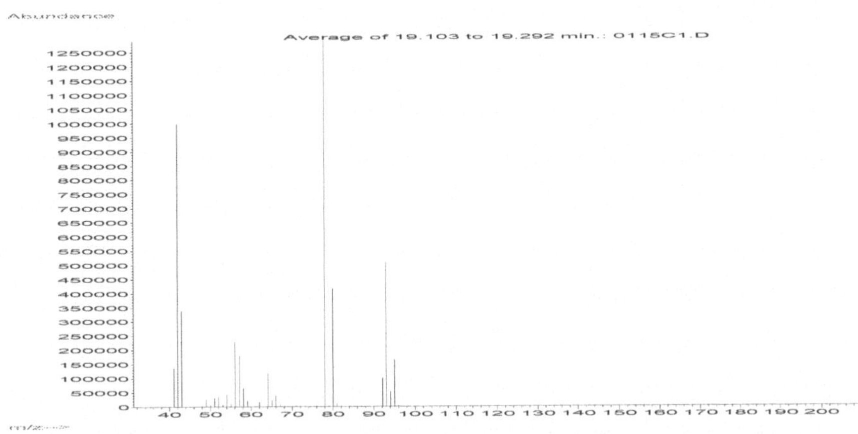

(b) Spectrum 2-4 Cl:N =0.925:1, w/o buffer, reaction time = 15 min

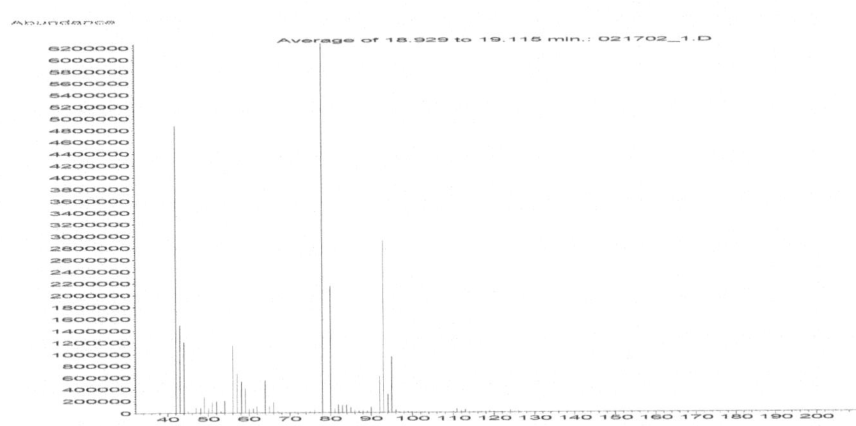

(c) Spectrum 2-5 Cl:N =10:1, w/o buffer, reaction time = 15 min

Figure **7-8. Chlorination of Ethylmethylamine at Different Cl:N Ratios.** (a) Cl:N=2, Reaction Time=6 Minutes; (b) Cl:N=0.925, Reaction Time = 15 Minutes; (c) Cl:N=10, Reaction Time=15 Minutes.

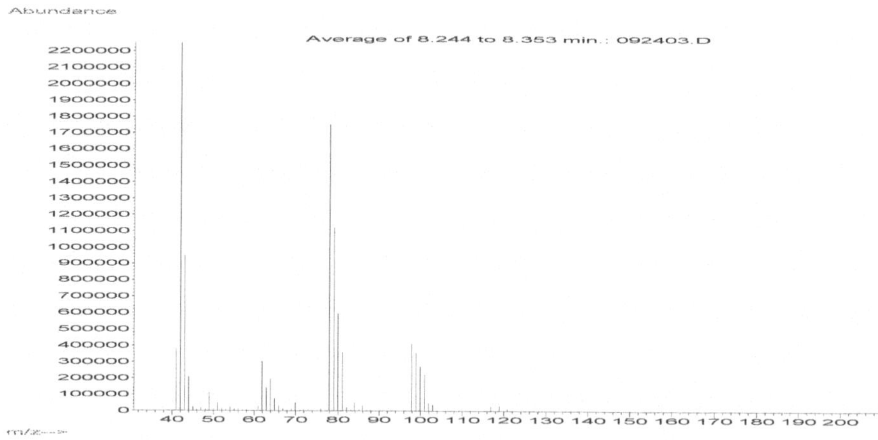

(a) Spectrum 3-3 Cl:N =2:1, pH = 9.20, reaction time = 4 min

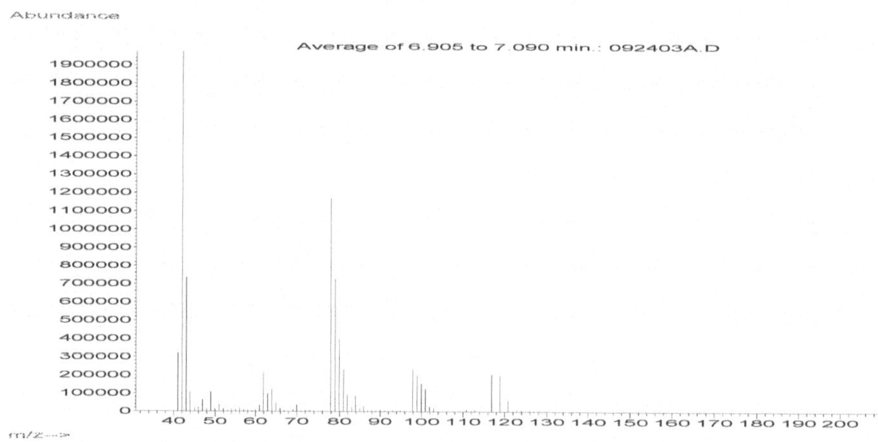

(b) Spectrum 3-3a Cl:N =2:1, pH = 9.20, reaction time = 300 min

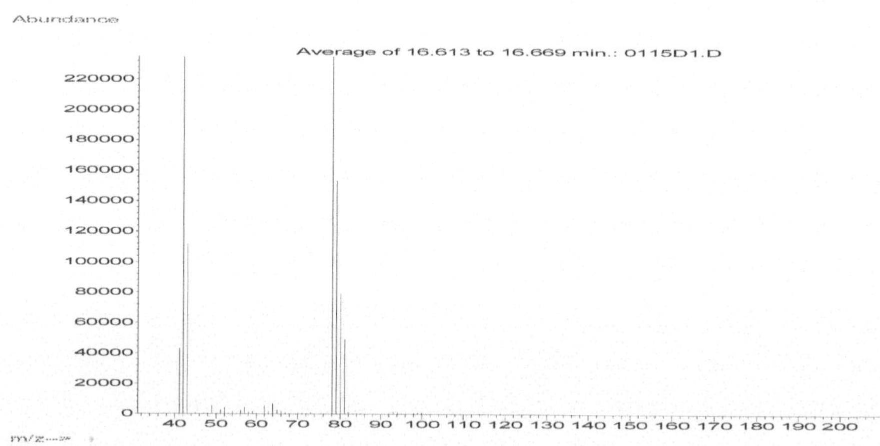

(c) Spectrum 3-4 Cl:N =0.925:1, w/o buffer, reaction time = 13 min

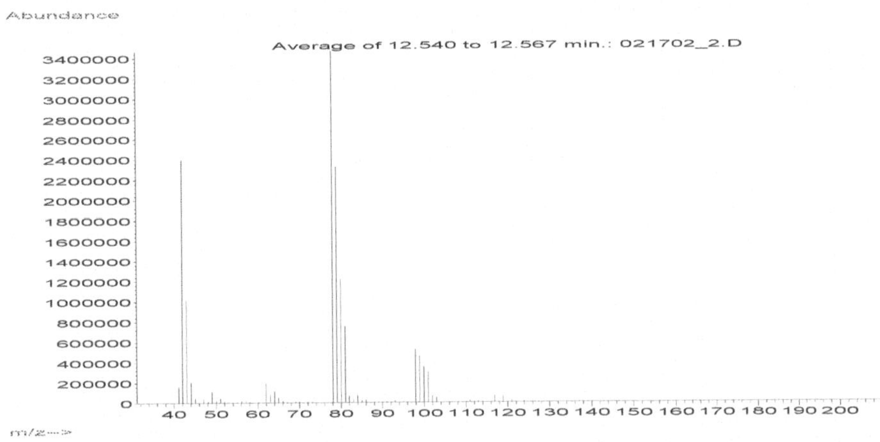

(d) Spectrum 3-5 Cl:N =10:1, w/o buffer, reaction time = 9 min

Figure 7-9. Chlorination of Trimethylamine at Different Cl:N Ratios. (a) Cl:N=2, Reaction Time = 4 Minutes; (b) Cl:N=2, Reaction Time=300 Minutes; (c) Cl:N=0.925, Reaction Time=13 Minutes; (d) Cl:N=10, Reaction Time=9 Minutes.

7.3.2 Toxicity Testing

The chlorination of aqueous solutions of methylamine, n-propylamine, ethylmethylamine, and trimethylamine by free chlorine was studied using MIMS. Since no other inorganic N was present in the sample, the only chlorination reagents in the sample were OCl^- and HOCl. This is relevant because free chlorine and inorganic chloramines both have the ability to transfer +1-valent chlorine to organic amines in aqueous solution.

The reaction monitoring experiments were conducted by continuously monitoring full mass spectra during the first 20 minutes of reaction. Chlorine transfer from free chlorine to aliphatic amine was rapid, with mass spectra that stabilized within five minutes. After chlorination, each organic chloramine was analyzed by MIMS at different reaction times to test product stability. The results showed a slight decay (<10%) in each organic chloramine after chlorination for 100 minutes. Figure 7-10 illustrates typical results for each organic chloramine from MIMS. Based on the isotope peaks, these spectra indicated that each compound contains one chlorine, since no evidence for dichlorinated products was observed from substitution reactions under the experimental conditions. Based on the MIMS results, the following reactions appear to describe the predominant behavior of these systems:

$$CH_3NH_2 + HOCl \leftrightarrows CH_3NHCl + H_2O \tag{7-1}$$
$$CH_3CH_2CH_2NH_2 + HOCl \leftrightarrows CH_3CH_2CH_2NHCl + H_2O \tag{7-2}$$
$$CH_3CH_2NHCH_3 + HOCl \leftrightarrows CH_3CH_2NClCH_3 + H_2O \tag{7-3}$$
$$N(CH_3)_3 + HOCl \leftrightarrows ClN(CH_3)_2 + CH_3OH \tag{7-4}$$

Further, a small peak of molecular ion (m/z=59) for each precursor was observed in those spectra, implying that there was essentially no precursor amine remaining after chlorination. DPD titration was used to determine the concentration of each organic chloramine and free chlorine concentration after reaction. Under the chlorination conditions, no free chlorine was detected after chlorination. Based on these results, it was assumed that each amine reacted with free chlorine completely and the resulting organic chloramines were the predominant compound present in solution.

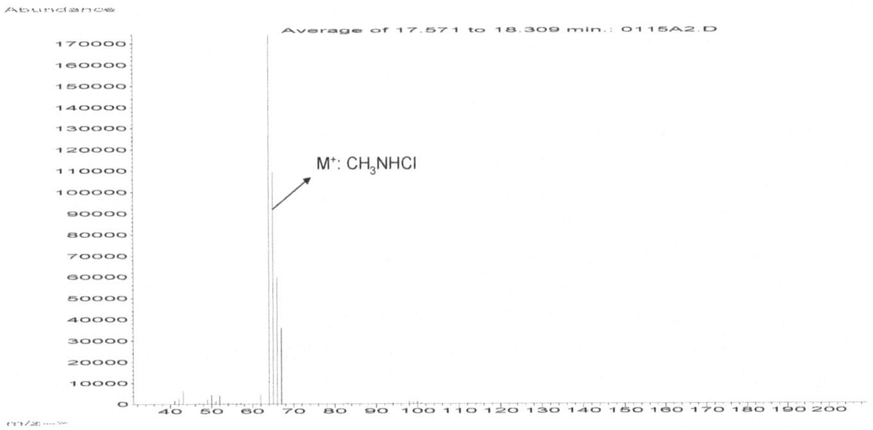
Abundance
Average of 17.571 to 18.309 min.: 0115A2.D
M+: CH3NHCl
m/z-->

(a)

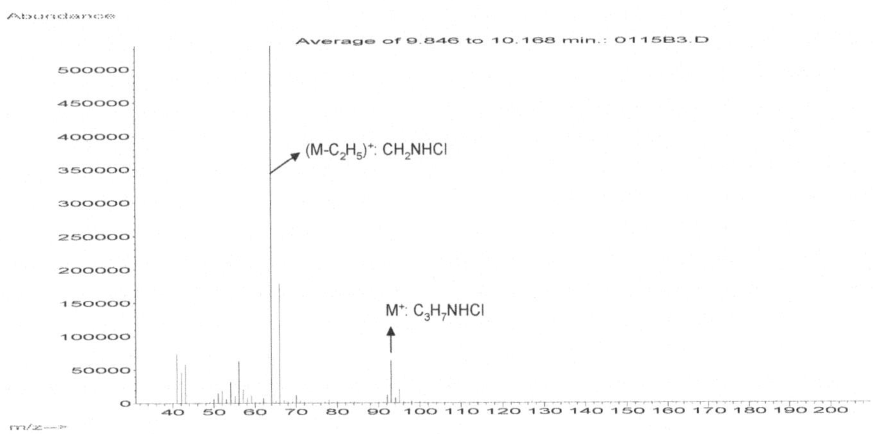
Abundance
Average of 9.846 to 10.168 min.: 0115B3.D
(M-C2H5)+: CH2NHCl
M+: C3H7NHCl
m/z-->

(b)

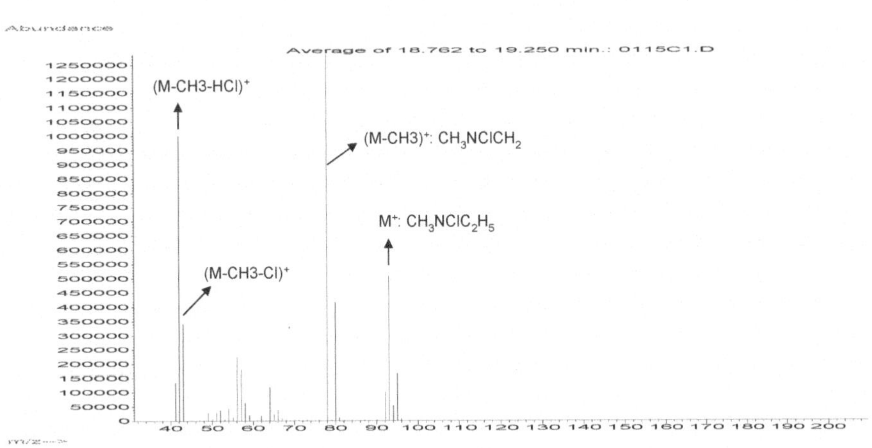
Abundance
Average of 18.762 to 19.250 min.: 0115C1.D
(M-CH3-HCl)+
(M-CH3)+: CH3NClCH2
M+: CH3NClC2H5
(M-CH3-Cl)+
m/z-->

(c)

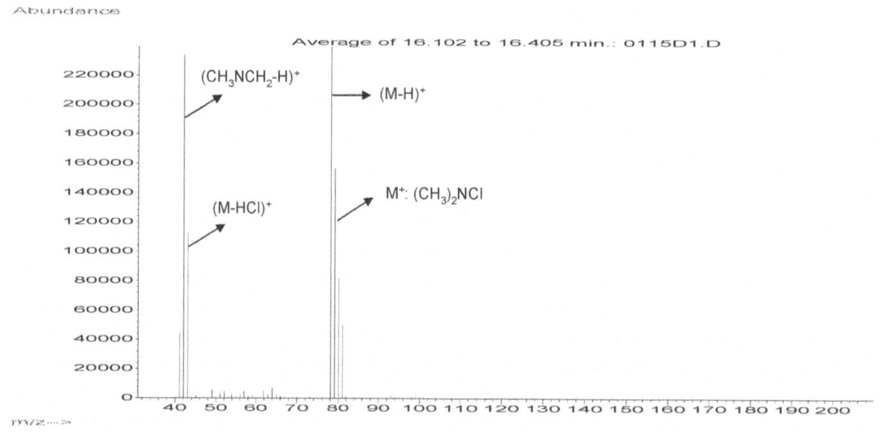

(d)

Figure 7-10. Typical Mass Spectra of Synthesized Organic Chloramines. (a) Chloromethylamine, $M^{+\cdot}$: $CH_3NH^{35}Cl$, $CH_3NH^{37}Cl$ (m/z=65, 67); (b) N-chloropropylamine, $M^{+\cdot}$: $C_3H_7NH^{35}Cl$, $C_3H_7NH^{37}Cl$ (m/z=93, 95), $(M\text{-}C_2H_5)^{+\cdot}$: $CH_2NH^{35}Cl$, $CH_2NH^{37}Cl$ (m/z = 64, 66); (c) chloroethylmethylamine, $M^{+\cdot}$: $CH_3N^{35}ClC_2H_5$, $CH_3N^{37}ClC_2H_5$ (m/z=93, 95), $(M\text{-}CH_3)^{+\cdot}$: $CH_3N^{35}ClCH_2$, $CH_3N^{37}ClCH_2$ (m/z=78, 80); (d) chlorodimethylamine, $M^{+\cdot}$: $(CH_3)_2N^{35}Cl$, $(CH_3)_2N^{37}Cl$ (m/z=79, 81), $(M\text{-}H)^{+}$ (m/z=78, 80).

7.3.2.1 Microtox

The Microtox assay was conducted on chlorinated and unchlorinated solutions of each precursor over a range of precursor concentrations. Analyses were performed in triplicate using exposure times of five and 15 minutes. The measured light levels were then converted to gamma ratio, defined as the ratio of the light decreased to the light remaining after the reagent is challenged by a sample. EC_{50} corresponds to the concentration that yields gamma ratio equal to 1. The acute toxicity responses of each precursor and product are presented as static five- and 15-minute EC_{50} values, which are defined as conditions in which 50% of the test microorganisms are affected due to the presence of the sample under designated exposure period.

The calculation of EC_{50} values was based on linear regression of applied concentration and the gamma values, both on a $\log_{10}$ scale. The gamma values used for calculation have been adjusted by control ratio (the ratio of light remaining to initial light level in blank sample) to account for the light loss in blank sample through the following equations:

$$\text{Gamma} = [(\frac{I_{tc}}{I_{oc}}) \times (\frac{I_0}{I_t})] - 1 \qquad (7\text{-}5)$$

Where, I_{0c}, I_{tc}: light intensity of control after exposure time of 0 and t, respectively.
I_0, I_t: light intensity of sample after exposure time of 0 and t, respectively.

Data with a very large (>100) or very small (<0.01) gamma value were considered invalid and rejected for regression calculation. The EC_{50} can be calculated by setting $\log_{10}$ gamma to 0 in the linear regression equation. The 95% CI of EC_{50} can be estimated statistically (Devore, 1995).

Figures 7-11 and 7-12 show the relationships between fraction of bacteria affected and applied concentration. Since there was essentially no free chlorine or precursor amine remaining in the chlorinated sample, the toxicity responses of amines after chlorination were assumed to be largely attributable to chlorinated amines. Table 7-4 summarizes the resulting EC_{50} values of each precursor and chlorinated amine. The EC_{50} value decreased (or toxicity response increased) by more than one order of magnitude after chlorination for each precursor amine.

The EC_{50} values for some common disinfection by-products on the same bacteria (*P. phosphoreum*) are higher than the organic amines used in this study. For example, the EC_{50} value of chloroform ranges from 3.62mM to 5.62 mM, depending on the exposure time (Elnabarawy et al., 1988; Qureshi et al., 1982). This implies that DBPs from organic-N compounds may be more toxic (toward some organisms) than some non-nitrogenous DBPs. In chlorination of wastewater containing high concentrations of organic-N compounds, this may be an important issue.

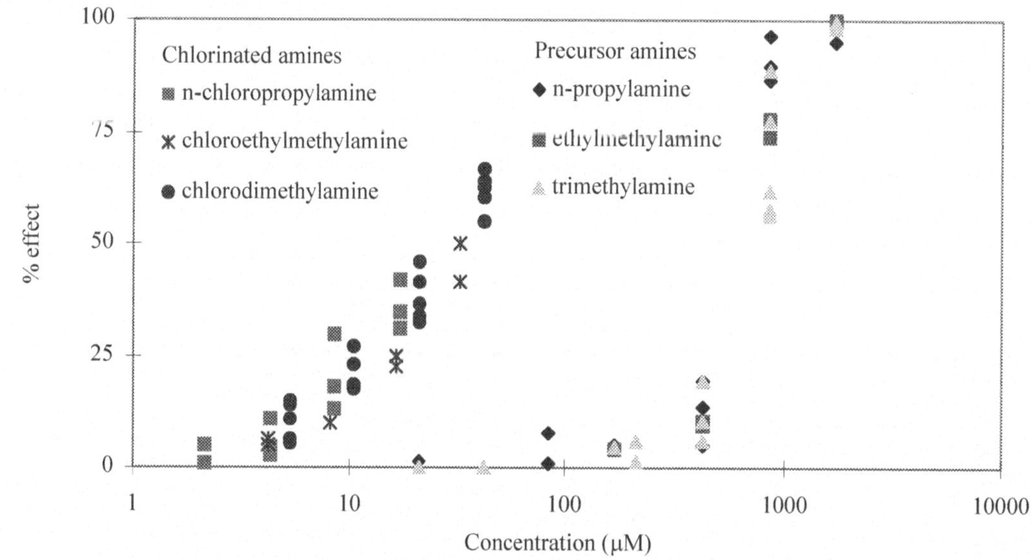

(a) Contact time = 5 minutes

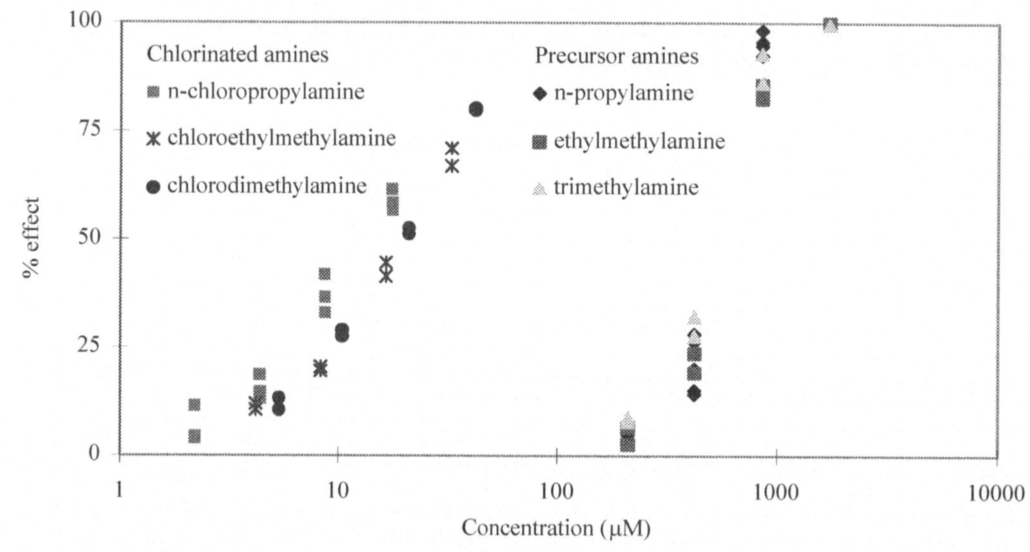

(b) Contact time = 15 minutes

Figure 7-11. Microtox Results of Precursor Amines and Chlorinated Amines (a) with Five-minute Contact Time; (b) with 15-minute Contact Time.

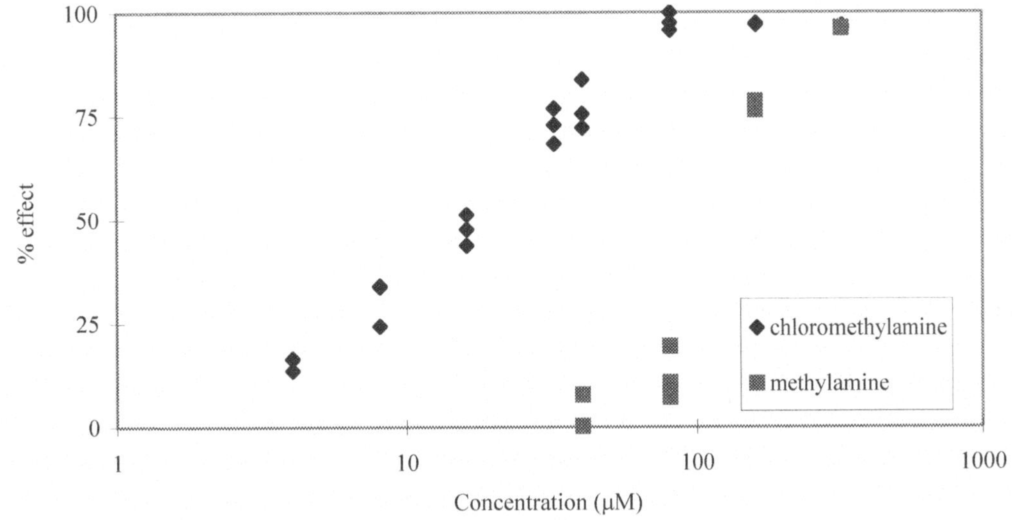

(a) Contact time = 5 minutes

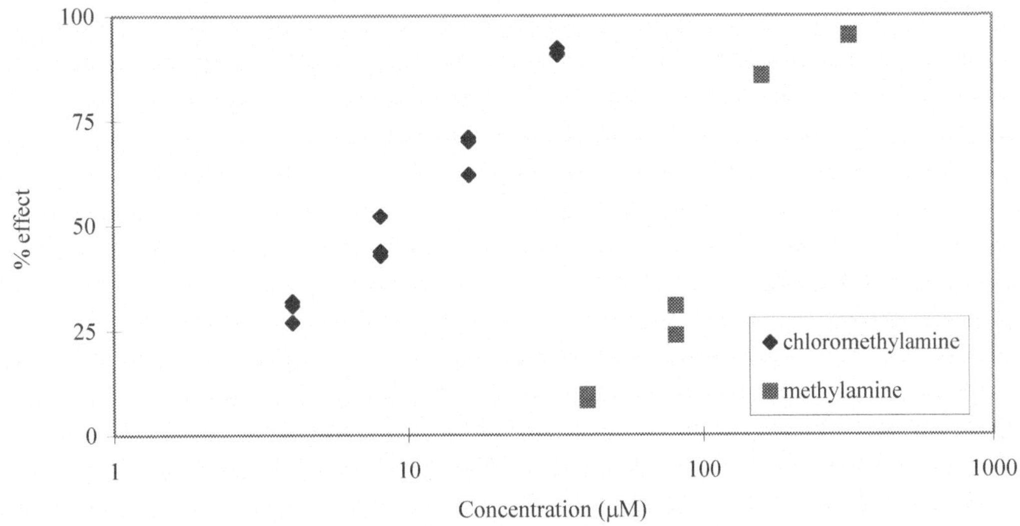

(b) Contact time = 15 minutes

Figure 7-12. Microtox Results of Methylamine and Chloromethylamine: (a) with Five-minute Contact Time (b) with 15-minute Contact Time.

Table 7-4. Summary of Microtox Results (Concentration in µM).

Compounds	5 Minutes		15 Minutes	
	EC_{50}	$CI_{95\%}$ of EC_{50}	EC_{50}	$CI_{95\%}$ of EC_{50}
Methylamine	120.74	104.00 – 140.39	99.43	89.51 – 110.44
Chloromethylamine	15.81	14.23 – 17.46	8.17	7.04 – 9.30
N propylamine	581.11	523.76 – 644.72	469.63	404.66 – 544.91
N-chloropropylamine	23.24	15.07 – 35.92	12.68	10.70 – 15.07
Ethylmethylamine	665.87	630.51 – 703.09	544.57	433.93 – 683.29
Chloroethylmethylamine	40.14	32.11 – 50.14	19.72	17.60 – 22.11
Trimethylamine	656.56	570.29 – 755.87	584.66	497.88 – 686.51
Chlorodimethylamine	29.86	25.92 – 34.23	15.92	14.51 – 17.61

7.3.2.2 Oxygen Utilization Rate Test

The OUR assay was used to characterize the inhibition of nitrifiers caused by the precursor amines and chlorinated amines. Nitrifiers are responsible for metabolizing nitrogenous materials obtained from sources such as activated sludge, soil, and sediment. Precursor amines and corresponding organic chloramines with concentrations equal to EC_{50} (from Microtox tests) were subjected to OUR tests, and no inhibition was observed (Figure 7-13). After these preliminary tests, higher concentrations of precursor amines and chlorinated amines were subjected to OUR tests, and the OUR of blank samples was measured before each test. The oxygen concentration of each sample was recorded with time, and the maximum OUR was calculated from the slope of the oxygen depletion curve. Figure 7-14 shows the resulting oxygen uptake rates for organic chloramines under low and high. Precursor amines, except ethylmethylamine, can cause respiratory inhibition of nitrifiers under the tested concentrations. After chlorination, chlorinated amines cause more inhibition in nitrifier respiration, and the inhibition increased with concentration of each chlorinated chloramine.

Table 7-5 summarizes the OUR estimates for each sample. Precursor amines cause inhibition to nitrifiers as compared against the blank and the OUR decreased by 41.6%, 12.2%, and 32.6% for n-propylamine (2.33 mM), ethylmethylamine (2.18 mM) and trimethylamine (1.28 mM), respectively. However, King and Painter (1985) suggested that for the inhibition of respiration test, it would be unwise to consider differences of less than one order of magnitude to be significant. The inhibition of each precursor amine increased after chlorination. Under the same concentration condition (0.82 mM), the decreases in OUR caused by organic chloramines were 44%, 70%, and 92.5% for n-chloropropylamine, ethylmethylamine, and trimethylamine, respectively. Ethylmethylamine caused little or no inhibition to nitrifiers; however, after chlorination, chloroethylmethylamine significantly decreased oxygen uptake rate under similar concentrations.

By comparing results from Microtox and OUR tests, researchers found that the Microtox system has higher sensitivity than OUR assays. Elnabarawy et al. (1988) compared three toxicity assays (Microtox, Polytox, and activated sludge) and found that Microtox was the most sensitive test to inorganic and organic compounds. They also found that the polytox and activated sludge results are comparable after correcting for biomass. There are potentially two toxicant removal mechanisms associated in these tests: adsorption and biodegradation. The relatively lower sensitivity of OUR assay may be attributable to higher adsorption because of its high MLSS concentration.

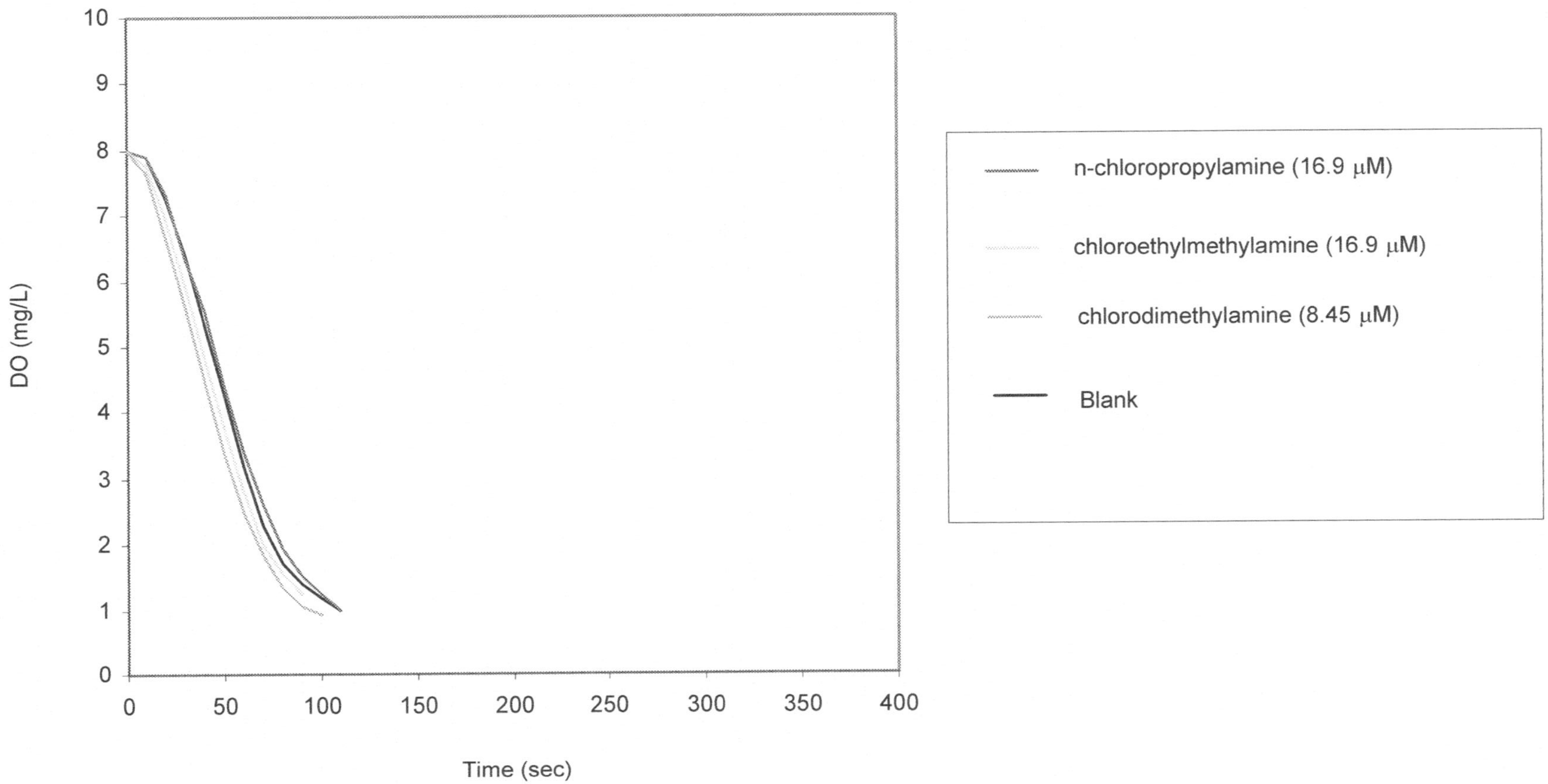

Figure 7-13. OUR Curves of Different Organic Chloramines Concentrations Relevant to EC_{50} from Microtox Test Results. No Significant Inhibition in Respiration Was Observed.

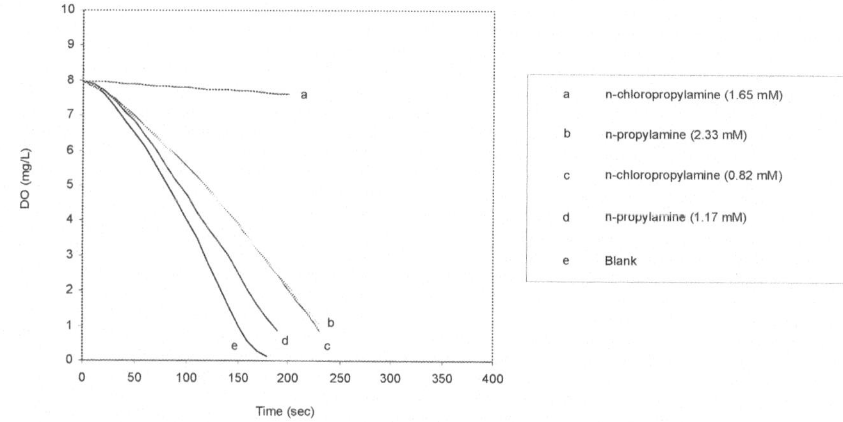

(a)

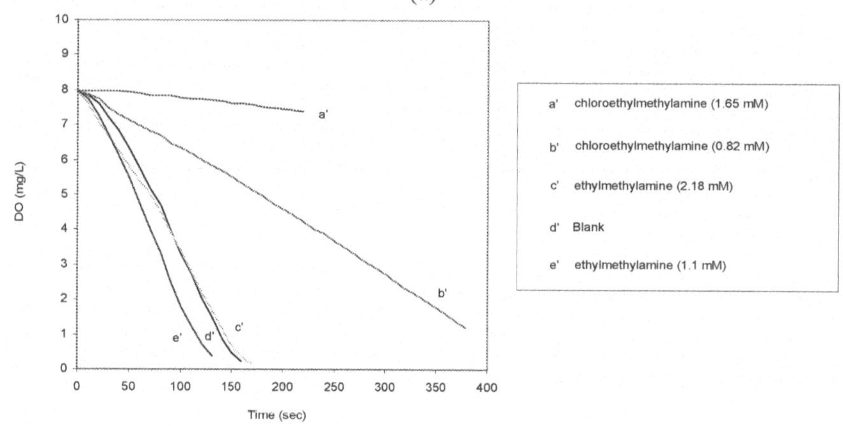

(b)

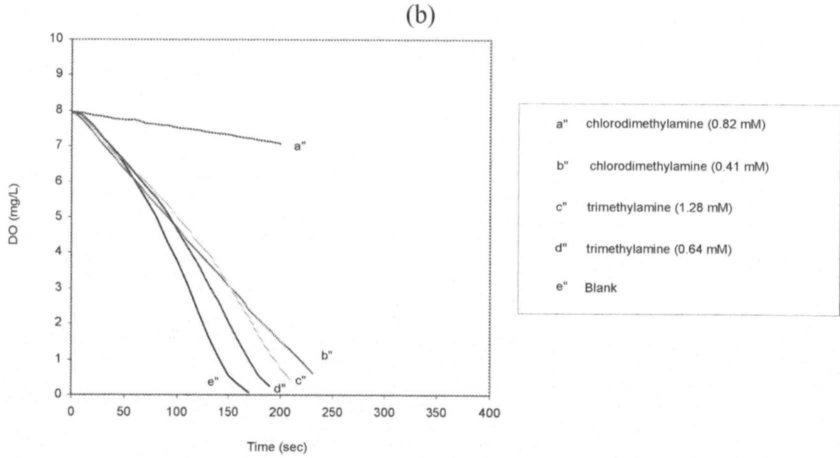

(c)

Figure 7-14. OUR Curves of Different Precursor Aliphatic Amines and Chloramines Under Different Concentrations: (a) N-propylamine and N-chloropropylamine; (b) Ethylmethylamine and Chloroethylmethylamine; (c) Trimethylamine and Chlorodimethylamine.

Table 7-5. Results of OUR Tests for Precursor Amines and Chlorinated Amines.

Sample	Concentration (mM)	Max. OUR (mg/LHr)	Inhibition (%)
N-propylamine	0	219	-
	1.17	159	27.4
	2.33	128	41.6
Ethylmethylamine	0	229	-
	1.10	259	-13.1
	2.18	201	12.2
Trimethylamine	0	242	-
	0.64	200	17.4
	1.28	163	32.6
Sample	**Concentration (mM)**	**Max. OUR (mg/LHr)**	**Inhibition (%)**
N-chloropropylamine	0	248	-
	0.82	139	44.0
	1.65	7	97.2
Chloroethylmethylamine	0	227	-
	0.82	68	70.0
	1.65	12	94.7
Chlorodimethylamine	0	227	-
	0.41	113	50.2
	0.82	17	92.5

7.4 Summary and Conclusions

Experiments were conducted to characterize the chemistry and toxicology of several simple aliphatic amines and their chlorinated forms. Chlorination of these compounds tends to lead to the production of chlorinated organic amines through Cl substitution for N-hydrogens; however, under some circumstances, Cl substitution for alkyl groups may also take place. The chlorinated compounds appear to be relatively stable in aqueous solution.

The simple aliphatic amines selected for study elicited relevant toxicological properties based on Microtox and OUR assays. Chlorination of these compounds, under conditions that are representative of many municipal wastewater chlorination systems, leads to the production of their N-chlorinated forms. These N-chlorinated compounds were substantially more toxic than their parent compounds (using both toxicity assays described above).

The results of these tests demonstrate the production of persistent chlorinated organic-N compounds in chlorination of model organic-N compounds. These data, together with previously published research, suggest that chlorination practices in municipal wastewater systems are likely to lead to the production of these same compounds, as well as many related chlorinated organic-N compounds. Of course, the concentration and toxicological significance of these compounds will depend on site-specific effluent characteristics. However, these results illustrate a potentially important limitation of chlorine-based disinfection systems for municipal wastewater treatment.

CHAPTER 8.0

MICROBIAL HEALTH RISK REDUCTION ASSOCIATED WITH WASTEWATER DISINFECTION

8.1 Background

Fecal contamination of water and exposure to untreated sewage has long been known to cause disease. The probability of illnesses any waterborne pathogen may cause will depend on the type of contact made, exposure, concentration of the organisms in the contaminated water, temperature of the receiving water, transport of the pathogen from source to contact point, and individual or population susceptibility to pathogen-borne illnesses (Meyland and Machlin, 2001). Exposure pathways for waterborne pathogens include shellfish consumption, skin contact or ingestion during recreation, direct contact, inhalation, and drinking water.

A waterborne disease outbreak occurs when two or more persons experience similar illness after consumption or use of a common water source, generally proven using epidemiologic methodologies (Rose et al., 2001a). Epidemiologic studies are not conducive to showing the full scale of waterborne disease outbreaks; no etiologic agents have been identified in half of the reported waterborne disease outbreaks in the United States (Enriquez et al., 1995). As few as 10% of outbreaks have been documented (Frost, 1996). Gastrointestinal (GI) illnesses are largely unreported due to the lesser severity of illness in healthy individuals. When etiologic agents have been identified, most often the source of the fecal contamination has not. Such is the case with wastewater discharges, where the data to address the association to health risks are generally sparse.

The purpose of this chapter is to examine and compare the probability of illness associated with exposure to non-disinfected wastewater effluents with those associated with wastewater effluents that had been subjected to UV irradiation or chlorination. The human exposure scenario employed was discharge of the wastewater to a recreational water body and exposure through swimming. While there are many microbial hazards in wastewater, this exercise focused on viruses.

8.2 Risk Assessment Framework

Risk assessment is the process of integrating scientific data regarding an environmental hazard into a framework to address the risk of exposure and the potential health impacts (Rose and Grimes, 2001). Through developing microbial risk assessments, scientists may identify potential health hazards, and they can discover the adequacy of the underlying data. A risk assessment permits scientists to focus on the data gaps that are most critical to improving the precision of their assessment.

The conventional microbial risk framework includes developing a quantitative database in four areas (Rose and Grimes, 2001):

- Hazard identification (types of pathogens and description of illnesses, hospitalization, and mortality),
- Dose-response (quantitative relationship between microbial exposure and outcome),

- Exposure assessment (prevalence, concentrations, distribution in time and space in water or food consumed), and
- Risk characterization (the quantitative likelihood of potential adverse health outcome based on the above).

For most waterborne pathogenic microorganisms, historical databases regarding occurrence in water or wastewater do not exist or are limited. For example, all of the microorganisms on the U.S. EPA's "Contaminant Candidate List" (CCL, Table 8.1) require a database for an assessment of exposure to be adequate. Of the 13 possible organisms on the CCL, nine are associated with municipal wastewater. Few of these organisms have been monitored in wastewater discharges. The contaminants included on the CCL are not subject to any current or proposed regulations. However, they are known or anticipated to occur in public water systems and are known or suspected to adversely affect public health, and therefore may require regulation under the Safe Drinking Water Act (SDWA; U.S. EPA, 2002). This list signifies a concern for the future, for which (at times) the risks may be unknown.

Table 8-1. Microbial Contaminant Candidate List (Gerba et al., 2003).

Category	U.S. EPA's Contaminant Candidate List
Bacteria	*Mycobacterium avium* complex ***Helicobacter pylori**** ***Aeromonas hydrophila**** *Legionella* (in groundwater)
Viruses	**Echoviruses*** **Caliciviruses*** **Hepatitis A virus*** **Adenoviruses***
Protozoa	***Cyclospora cayetanensis**** **Microsporidia*** ***Toxoplasma gondii****
Other	Blue Green Algae, Cyanobacteria *Acanthamoeba* (guidance for contact lens wearers)

* Of fecal origin

Waterborne disease statistics and potential risks are commonly divided into recreational outbreaks and drinking water outbreaks. Foodborne outbreaks can also be associated with contaminated water, but will not be addressed here. Current data regarding waterborne disease in the United States shows that during the last few years of reporting there has been an increase in disease associated with drinking water and recreational water (Lee et al., 2002). Many U.S. outbreaks go unreported, as the surveillance/reporting system is passive. Also, in most outbreaks the microbial pathogen is not identified. During 1995 to 1998, 11,612 laboratory-confirmed cases of cryptosporidiosis were reported to the Center for Disease Control (CDC) (Dietz and Roberts, 2000). Yet, recent estimates suggest that up to 300,000 cryptosporidiosis infections occur each year in the U.S.; many infected individuals do not seek medical attention (Dietz and Roberts, 2000). The most identifiable causes of gastroenteritis are bacteria, viral pathogens, and then parasites.

Municipal wastewater containing microorganisms may enter waterways via rainfall events. During 1996 nearly 3,700 beach closings and advisories were issued at U.S. ocean, bay, and Great Lakes beaches, with the detection of excessive concentrations of fecal bacteria causing 83% of the closings, often due to rainfall (Rose et al., 2001). The numbers of closures as a result of contaminated water demonstrates potential public health risks associated with recreational water. The bacterium *Shigella* is one of the leading causes of recreational waterborne outbreaks in lakes and rivers. *Shigella* is associated solely with human feces. This exemplifies how municipal wastewater can be implicated in waterborne disease outbreaks.

Human enteric viruses are associated solely with human wastes, and numerous outbreaks have been attributed to viruses following exposure to fecally polluted waters or following the consumption of shellfish in such waters (Tree et al., 2003). The concentration of viruses in a water body increases with the concentration of suspended particles, and even low concentrations can be sufficient to supply an infectious dose, resulting in a full-blown viral infection (U.S. EPA, 2002).

One of the most important parasites associated with waterborne disease is *Cryptosporidium.* This pathogen is associated with animal and human waste and was first detected as a cause of illness in humans in 1976. Since that time, it has been well recognized as a cause of diarrheal illness. It is extremely resistant to chlorination and has been found in surface waters throughout the United States and the world (Rose, 1997; Smith and Rose, 1998). The largest outbreak in the United States occurred in Milwaukee, Wisconsin, in 1993, where roughly 400,000 people became ill and over 100 died due to contamination of the water supply. Municipal wastewater has been implicated as the source of the contamination (MacKenzie et al., 1994; Rose et al., 2002).

During the reauthorization of the SDWA, sensitive populations, including the elderly, immunocompromised individuals, pregnant women, and infants, were considered in terms of their risk of waterborne disease. Literature indicated that this segment of the population was at relatively high risk of serious outcomes, including hospitalization and death. This has proven to be true with recent outbreaks of *E. coli* 0157H7 and *Cryptosporidium* where elderly, young, and the immunocompromised individuals died as a result of the outbreak (Gerba et al., 1996). These groups currently represent almost 20% of the population in the United States and are expected to greatly increase by the beginning of the next century, because of extended life spans and the increasing number of immunocompromised individuals. Within developed countries, more than half of the illnesses due to gastroenteritis and hepatitis A occur in the elderly population. Pregnant women have a case fatality ratio from hepatitis E infections during waterborne disease outbreaks ten times greater than the general population. Enteric diseases have the greatest adverse effects on immunocompromised individuals, such as AIDS and cancer patients.

8.3 The Microbial Hazards—Pathogens in Municipal Wastewater

Pathogens are found in varying concentrations in wastewater, whether it receives no treatment, primary treatment, or secondary treatment. Tables 8-2 through 8-4 describe the bacterial, protozoan, and viral pathogens that may be found in untreated municipal wastewater and the diseases they have the ability to cause (Rose and Grimes, 2001), including diseases attributed to the fecal-oral route via human feces. Hundreds of bacteria, parasites, and viruses can be found in untreated wastewater. In the U.S. there is no program to monitor for these pathogens in wastewater discharges.

Table 8-2. Some Pathogenic Bacteria Found in Wastewater.

Species	Disease(s)	Waste(s)
Aeromonas hydrophila	Septicemia, wound infections, diarrhea	Feces, domestic sewage, ambient waters (fresh and estuarine)
Aeromonas sobria	Septicemia, wound infections, diarrhea	Feces, domestic sewage, ambient waters (fresh and estuarine)
Aeromonas caviae	Septicemia, wound infections, diarrhea	Feces, domestic sewage, ambient waters (fresh and estuarine)
Bacteroides melaninogenicus	Orofacial	Human mouth and feces
Clostridium difficile	pseudomembranous colitis	Vagina and GI tract
Clostridium perfringens	Gangrene, wound abscesses, food poisoning	Human feces
Enterobacter spp.	Nosocomial	Domestic sewage
E. coli 0157: H7	Bloody diarrhea, uremic hemylolytic syndrome	Domestic sewage
Escherichia coli	Gastroenteritis	Domestic sewage
Fusobacterium necrophorum	Liver and soft tissue abscesses	Domestic sewage, animal feces
Klebsiella pheumoniae	Pneumonia, bacteremia, nosocomial	Water, feces, soil, plants
Morganella morganii	Urinary tract, nosocomial	Water, feces, decaying animals
Plesiomonas shigelloides	Gastroenteritis	Water, fish, aquatic animals
Proteus spp.	Urinary tract, nosocomial	Water, feces, decaying animals
Pseudomonas aeruginosa	Burn, wound, corneal, ear, urinary, lung, skin, GI tract	Water, domestic sewage, plants, sediment, fish
Salmonella typhi	Typhoid fever	Domestic sewage
Salmonella enteritidis	Gastroenteritis and septicemia	Domestic sewage, animal wastes, food, compost
Shigella boydii	Bacillary dysentery	Human feces, domestic sewage
Shigella dysenteriae	Bacillary dysentery	Human feces, domestic sewage
Shigella flexneri	Bacillary dysentery	Human feces, domestic sewage
Shigella sonnei	Bacillary dysentery	Human feces, domestic sewage
Vibrio cholerae	Asiatic cholera	Domestic sewage, shellfish, saltwater
Yersina enterocolitica	GI, acture mesenteric lymphadenitis	Water, milk, mammalian alimentary canal

Table 8-3. Human Enteric Viruses Found in Wastewater.

Virus	Disease(s)	Waste(s)
Adenoviruses		
Human adenovirus	Acute respiratory, pharyngitis, acute hemorrhagic cystitis	Domestic sewage
Enteric adenovirus	Gastroenteritis	Domestic sewage
Astrovirus	Gastroenteritis	Domestic sewage
Calciruses		
Calcivirus	Gastroenteritis	Domestic sewage
Noroviruses	Gastroenteritis	Domestic sewage
Coronaviruses		
Enteric coronavirus	Intestinal disorders	Uncertain
Orthomyxoviruses		
Parvovirus	Gastroenteritis	Domestic sewage
Picornaviruses		
Coxsackievirus A	Meningitis, herpangia, common cold	Domestic sewage
Coxsackievirus B	Mycarditis, pleurodynia, rash, meningitis, paralysis	Domestic sewage
ECHO virus	Paralysis, diarrhea, meningitis	Domestic sewage
Hepatitis A virus	Infectious hepatitis	Domestic sewage
Poliovirus	Poliomyelitis	Domestic sewage
Reoviruses		
Reovirus	Respiratory, gastroenteritis	Domestic sewage
Rotavirus	Infantile diarrhea	Domestic sewage

Table 8-4. Waste Associated Protozoa Pathogenic for Humans.

Species	*Disease(s)*	*Waste(s)*
Mastigophora (flagellates)		
Giardia lamblia	Diarrhea	Human feces
Sarcodina (amebas)		
Entamoeba histolytica	Amebic dysentery	Human and other animal feces, domestic sewage
Dientamoeba fragilis	Mild diarrhea	Human feces
Sporozoa		
Cryptosporidium spp.	Diarrhea	Human and animal feces
Sarcocystis spp.	Sarcocystosis	Animal feces
Ciliata		
Balantidium coli	Balantidiasis	Animal feces, especially swine

The bacteria *E. coli* (including 0157:H7), S*almonella* species, and *Shigella* species remain important causes of disease in the U.S. All the viruses in Table 8-3 are known to cause disease in the U.S. Enteric viruses come only from human wastes. They can cause acute and chronic disease that affect the GI tract, liver, heart, and meninges. Adenoviruses, Calciviruses, Picornaviruses, and Rotaviruses cause hundreds of thousands of cases of disease per year. The exception is Poliovirus, as the vaccination program has reduced the number of infections. The enteroviruses include Coxsackievirus which are on the CCL. They can cause many types of disease and can be often detected in sewage contaminated water. Table 8-4 includes the parasites *Cryptosporidium* and *G. lamblia*, which cause the greatest numbers of cases in the U.S. These are enteric protozoa; they produce oocysts and cysts which are resistant to many water treatment processes and are persistent in the environment.

8.4 Dose-Response

Dose-response relationships describe the association between the extent of the exposure (dose) to a pathogen and the occurrence of related effects. These models represent methods for characterizing pathogen infectivity. Studies have been conducted wherein numerous subjects (humans) were given a range of doses and the outcomes were determined in each group at each dose. Outcomes in these tests are synonymous with colonization, which defines infection, disease, and/or symptoms or antibody response in the exposed individual. A relationship between the dose administered and probability of infection or disease in the exposed population can be modeled mathematically. These models have been used in assessing drinking water risks and in the development of the *Surface Water Treatment Rule* (Regli et al., 1991; Haas et al., 1999).

The selection of the most appropriate risk model can be a challenging task. The Beta Poisson model or the exponential model have been used most frequently and are used to predict a no threshold effect. A no threshold effect involves predicting the potential for a single organism to initiate an infection. The exponential model is defined as follows:

$$P = 1 - \exp(-rN)$$

where N is the exposure, as measured in number of organisms and (r) equals a constant associated with the organism's probability of causing infection. The Beta Poisson model recognizes that there are variations in the probability of pathogen-host survival (r factor in the exponential formulation). Variations exist within the microbial population, such as the ability to survive, the ability to transcend the host defenses, and variations in the host's response to pathogen incursion. The Beta Poisson model formulation is:

$$P_i = 1 - \left(1 + \frac{N}{\beta}\right)^{-\alpha}$$

Alpha and beta are organism-specific parameters that define this variability in the dose response curve. P represents the probability of infection from single-dose exposure. N is the number of organisms ingested. In the application of the Beta Poisson model, the parameter N_{50} is commonly used. N_{50} represents the number of organisms required to infect 50% of the population. Table 8-5 describes the infectivity of several pathogens based on these two models. The data on dose-response indicate that Rotavirus is the most infectious wastewater pathogen, followed by *Giardia*, *Cryptosporidium*, and then the bacteria. These data were (for the most part) developed in humans and thus are suitable for risk assessment without extrapolation between animal and humans. The models have been used in addressing parasite and virus risks for drinking water and for regulatory development.

Table 8-5. Reported Dose-response Models for Waterborne Pathogens.

Microorganism	Model	Animal	Reference
Campylobacter	Beta-poisson α=0.145 N_{50} = 896	Humans	Haas et al., 1999
Cryptosporidium	Exponential r= 0.0042	Humans	Haas et al. 1996
E. coli	Beta-poisson α=0.1952 $N_{50} = 3.01 \times 10^7$	Humans (included enteroinvasive, enterotoxigenic, and enteropathogenic strains)	Haas et al., 1999
E.coli 0157 H7	Beta-poisson α=0.49 $N_{50} = 5.96 \times 10^5$	Infant rabbits	Haas et al. 2000.
Giardia	Exponential r=0.01	Humans	Rose et al., 1991
Salmonella	Beta-poisson α=0.3126 $N_{50} = 2.36 \times 10^4$	Humans (non-typhoid, multiple strains)	Haas et al. 1999
Shigella	Beta-poisson α=0.21 $N_{50} = 1.12 \times 10^3$	Human (*flexnerri* and *dysenteriae* pooled)	Haas et al., 1999
Rotavirus	Beta-poisson α=0.2531 N_{50} = 6.17	Human	Haas et al., 1993, Gerba et al., 1996.

8.5 Exposures: Pathogen Occurrence in Wastewater

In North America, secondary treatment generally includes a biochemical treatment operation (e.g., activated sludge) that accomplishes removal of BOD, separation of solids, and possibly oxidation of ammonia-N. Effluent quality is highly variable among wastewater treatment facilities due to their varying designs and operations. Secondary treatment typically removes roughly 85% of the organic matter in municipal wastewater by making use of the bacteria in it (U.S. EPA, 1998). However, Koivunen et al. (2003) demonstrated significant numbers of bacteria remained in secondary effluent after efficient removal of suspended solids and organic load. Primary treatment typically precedes secondary treatment, although stabilization ponds and aerated lagoons commonly forego dedicated primary treatment (National Research Council [NRC], 1998).

The typical microbial community within activated sludge, consisting of bacteria, protozoa, and metazoan, is vital for the removal of viruses and for a significant reduction in *E. coli* 0:157-phage (Tanji et al., 2003). Chauret et al. (1999) showed that during aerobic wastewater treatment, *Cryptosporidium* and *Giardia* were reduced by 2.96 $\log_{10}$ and 1.40 $\log_{10}$, respectively. *C. perfringens* spores, *C. perfringens* total counts, somatic coliphages, and heterotrophic bacteria were reduced by 0.89 $\log_{10}$ to 2.02 $\log_{10}$ units. All other microorganisms quantified in this study were reduced by at least 3.53 $\log_{10}$ units (Chauret et al., 1999). Table 8-6 lists typical microorganism removal efficiencies for common wastewater treatment processes. To achieve levels of acceptable risk the removal of pathogens from municipal wastewater that impact drinking water supplies may need to be as great as 6 $\log_{10}$ to 8 $\log_{10}$ units. This is due to the high concentrations of pathogenic microorganisms that can be found in municipal wastewater (Table 8.7). Several orders of magnitude greater removal (4 $\log_{10}$ to 5 $\log_{10}$ units greater) than what is currently achieved (Table 8-7) may be needed to accomplish acceptable risk in situations where drinking water supplies are directly influenced by municipal wastewater effluent discharges.

Table 8-6. Typical Percentage Removal of Microorganisms by Conventional Treatment Processes.

Microorganism	Primary Treatment	Secondary Treatment	
		Activated Sludge	**Trickling Filter**
Fecal coliforms	<10	0-99	85-99
Salmonella	0-15	70-99	85-99
Mycobacterium tuberculosis	40-60	5-90	65-99
Shigella	15	80-90	85-99+
Entamoeba histolytica	0-50	Limited	Limited
Helminth ova	50-98	Limited	60-75
Enteric viruses	Limited	75-99	0-85

Source: NRC, 1998.

Table 8-7. Reported Concentrations of Pathogenic and Indicator Microorganisms in Secondary Effluent from WWTPs. Average Levels Reported (CFU, PFU, or cysts/oocysts per 100 L).

Pathogens	Occoquan, Virginia	St. Petersburg, Florida	Tampa, Florida	California (Data from 3 facilities)
Clostridium	4452	NT	NT	NT
Total Coliforms	170,000	1.5×10^6	NT	NT
Fecal Coliforms	7764	190,000	NT	NT
Enterococci	2186	NT	NT	NT
Coliphage	1821	5×10^5	NT	NT
Enterovirus	75.8	20	3.5	650 55 56
Cryptosporidium Oocysts	Not detected	140	NT	NT
Giardia Cysts	2297	440	5	NT

Source: NRC, 1998
NT, not tested

Cryptosporidium oocysts are generally more resistant to water treatment processes and disinfection practices than other ubiquitous waterborne microorganisms and are common in surface water bodies. Plants that use sand filtration along with activated sludge have shown significantly lower levels of oocysts in finished effluents than those using activated sludge treatments alone (Bonadonna et al., 2002). *Cryptosporidium* oocysts have been reported in secondary effluent at a concentration of 140 oocysts/100 L, while *Giardia* cysts were found to range from 440 cysts/100 L to 2297 cysts/100 L (NRC, 1998).

8.6 Disinfection

Chlorination and UV irradiation are the most prevalent forms of wastewater disinfection in the United States. The effectiveness of UV is largely dependent on the applied UV dose, suspended solids content, UV transmittance, non-disinfected microbial concentration, and the degree of association of microorganisms with particles (George et al., 2002). Linden et al. (2002) demonstrated that UV disinfection is effective in the inactivation of *G. lamblia* cysts in water with low turbidity. Jacangelo et al. (2003) examined removal and inactivation of enteric pathogens and other microbial indicators through UV disinfection following secondary treatment. The UV dose required to achieve reductions in the concentration of indigenous F-specific bacteriophage was approximately 20 mJ/cm^2 per $\log_{10}$ unit removal. The concentration of enterovirus and adenovirus were reduced to the limit of detection at UV doses of 35 mJ/cm^2 to 40 mJ/cm^2 and 40 mJ/cm^2 to 45 mJ/cm^2, respectively. *C. perfringens* spores were the most resistant indicator organisms, requiring a UV dose of 75 mJ/cm^2 (Jacangelo et al., 2003).

Done properly, chlorination following secondary treatment will inactivate >99% of the harmful bacteria in effluent (U.S. EPA, 1998). Viruses and parasites found in municipal wastewater, whether primary or secondary, are characterized as being much more resistant and have differing sensitivities to chlorination. There are currently no data to demonstrate that *G. lamblia* cysts are inactivated during chlorine-based disinfection of secondary effluents. Recent studies on infectivity of *Cryptosporidium* have found no inactivation due to chlorination of even highly treated wastewaters (Gennacarro et al., 2003).

Chlorine disinfection can inactivate some viruses in wastewater, but not as effectively as it does in drinking water because of interference by dissolved organics and suspended particulates; unless ammoniacal nitrogen is removed from wastewater, the predominant form of chlorine will be (inorganic and organic) chloramines, which are generally regarded as being less effective against viruses and parasites than free chlorine (Mahin and Pancorbo, 1999). The efficiency of chlorine-based disinfection processes is also known to be affected by temperature and pH (Gerba et al., 2003; Sobsey et al., 1989).

Tree et al. (2003) examined primary treatment of sewage coupled with chlorine disinfection and demonstrated that enterococci were more resistant to chlorination than *E. coli*; both bacteria were more rapidly inactivated than the viruses examined. Poliovirus, compared to the bacterial indicators, was more resistant to chlorination. FRNA bacteriophage was the most resistant organism to disinfection by chlorine. The study indicated that neither *E. coli* nor enterococci were sufficient indicators of virus removal during primary sewage disinfection.

8.7 Analysis of Risk

One may use quantitative risk assessment models to extrapolate the risk of exposure in pathogen-contaminated waters associated with municipal wastewater discharges. To accomplish this task, it is generally necessary to use published data regarding pathogen concentrations, as well as reported values of removal or inactivation achieved by disinfection. These data are incorporated into an appropriate probability of infection model and analyzed in the context of a specific exposure scenario. The lack of epidemiological data and active assessment makes it virtually impossible to examine the numbers of sporadic cases of disease associated with wastewater discharges; however, it is possible to examine the practice using a quantitative microbial risk assessment approach.

In order to examine the risks in this study, the following scenario was used: the hazard, dose-response and exposure were focused on enteric viruses for this analysis. There were a number of reasons for electing to focus on viruses (Haas et al. 1996; Rose et al., 1991; Gerba et al. 1996):

- Viruses are more resistant to chloramination than the coliform bacteria and are one of the most resistant targets for UV disinfection:
- Viruses have a low infectious dose and represent a range of illnesses.
- Viruses are used as a target organism for designing disinfection systems in some applications. For example, California Title 22 is focused on virus inactivation.
- The dose-response function for rotaviruses has been used in drinking water risk assessments (Regli et al., 1991)
- Concentrations of viruses can be estimated in the five facilities from which samples were analyzed in this research without disinfection, with UV disinfection, and with chlorination based on the published literature and the results of viral inactivation studies described in Chapter 6.0.

The exposure scenario used for these calculations was a single swimming event with 100 mL ingestion. The virus concentration used in this exercise was 75.6 PFU/100 L as reported by Rose et al. (2001) as an average value in a 12-month study of a full-scale facility's secondary effluent. This virus concentration was combined with the measured $\log_{10}$ reductions achieved via UV disinfection and chlorination in the bench-scale (batch) exposures studies indigenous coliphage (see Chapter 6.0). Since the measured values at the relevant doses appear to be at or below the limit of detection, the measured inactivation rate constants for UV and chlorine-based disinfection were used to estimate the viral concentration in the disinfected effluent. Virus inactivation by disinfection ranged from 0.32 $\log_{10}$ to 3.61 $\log_{10}$ units and were generally greater when using UV than with chlorine. Facilities A, B, and D achieved the greatest reductions via UV, while Facilities C and E achieved greater or equivalent virus reductions by use of chlorine (Table 8-8). Water quality characteristics in each facility likely impacted the virus inactivation and the inactivation was also dependent on the type of bacterial host used, with the F+amp host generally restricted more to analyzing the RNA viruses which have been found to be more resistant to disinfection.

Table 8-8. Average $\log_{10}$ Reductions achieved for Coliphage During Disinfection of Secondary Effluent by UV Irradiation and Chlorination. Values for Each Facility were Extracted from Data Presented in Chapter 6.0.

Facility Identifier (phage host)	Log_{10} Reductions of Viruses				
	UV Dose (mJ/cm^2)[a]			Chlorine contact time (min)[b]	
	5	**10**	**20**	**20**	**40**
E (*E.coli*) (F+amp)	047 0.68	0.94 1.37	1.88 2.74	1.81	3.61
B (*E.coli*) (F+amp)	0.88 0.59	1.75 1.19	3.51 2.38	0.25 0.13	0.5 0.26
C (*E.coli*)	0.42	0.84	1.69	0.78	1.56
D (*E.coli*) (F+amp)	0.69	1.37	2.74	0.32 0.43	0.64 0.87
A (*E.coli*) (F+amp)	0.64 0.36	1.27 0.73	2.54 1.45	0.3 0.26	0.59 0.52
Mean ***E.coli*** **F+amp**	**0.61** **0.49**	**1.21** **0.99**	**2.42** **1.97**	**0.37** **0.25**	**0.75** **0.5**

[a]Exposure conducted in a well-mixed batch reactor under a collimated beam.
[b]Exposure conducted in a well-mixed batch reactor with an initial chlorine concentration of 2.0 mg/L (as Cl_2).

Figure 8-1 provides a schematic representation of the process used to estimate the risk of infection based on the exposure scenario defined above. The F+amp data were used except in the case where there was only *E.coli* host data. In the case of UV disinfection, only the data corresponding to doses of 10 and 20 mJ/cm^2 were used, as these doses were shown in other experiments to be representative of UV exposure scenarios to be applied in municipal wastewater treatment facilities. Estimated risks associated with these simulations are summarized for each wastewater effluent examined in this research in Table 8-9. The risk without disinfection was 4.4×10^{-3}. Risks after disinfection ranged from 1.3×10^{-3} to 1.2×10^{-6}. The effluent from Facility A yielded the highest risk to downstream users when using chlorine. UV disinfection of the effluents from Faciliites D and E yielded the lowest risk, with UV dosing at 10 mJ/cm^2 yielding a risk of 1.8 x 10^{-4}, while a UV dose of 20 mJ/cm^2 yielded a risk lower than 8.1 x 10^{-6}. Regardless of contact time, chlorination generally resulted in an effluent that yielded a greater risk of infection than with UV irradiation. The one exception to this generalization was the effluent from the Chicago facility, where chlorination and UV irradiation achieved equivalent risk reductions.

In examining the data in Tables 8-8 and 8-9, it is important to reiterate that the disinfection exposure scenarios applied in this research were chosen to yield effluent composition that was in compliance with existing effluent discharge permit limitations at each facility. Therefore, the criteria for design of these disinfection exposures did not include a reduction of risk of exposure to enteric viruses. The risk calculations presented in this chapter are representative of the reductions in risk associated with current disinfection practices.

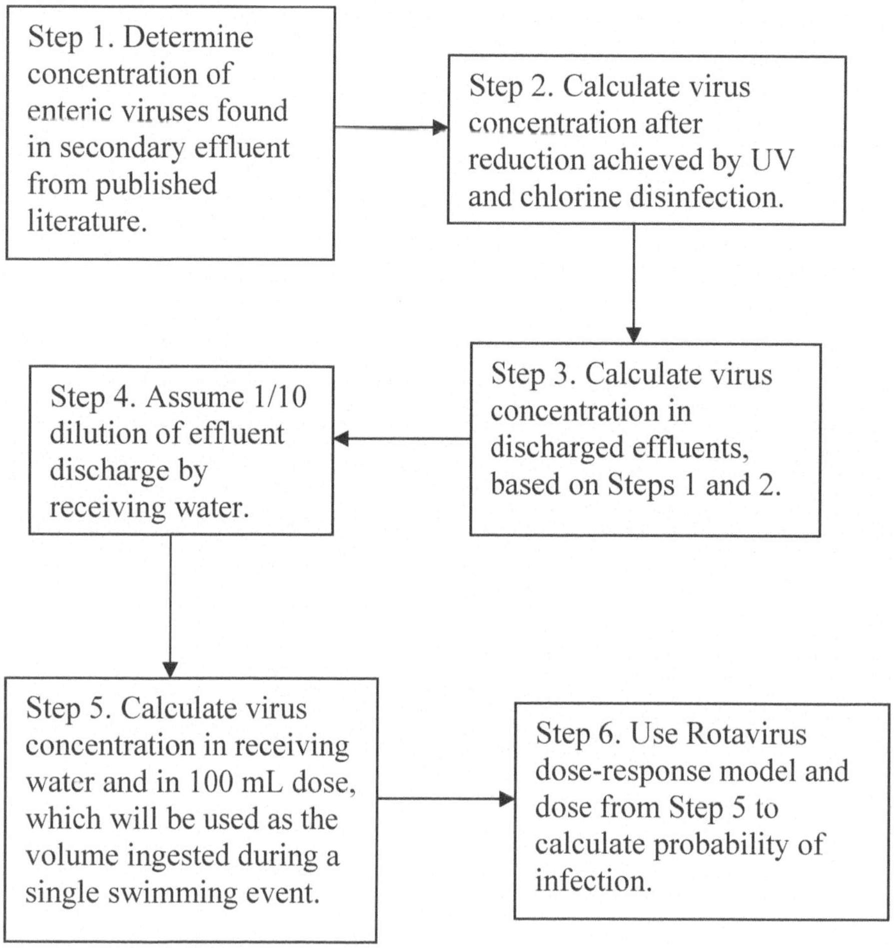

Figure 8-1. Approach Used for Calculating the Risk of Infection.

Table 8-9. Probability of Virus Infection from Swimming in Waters Receiving Disinfected Wastewater.

	Probability of Infection			
	UV Dose (mJ/cm^2)[a]		Chlorine contact time (min)[b]	
Facility Identifier	10	20	20	40
E	$1.88x10^{-4}$	$8.12x10^{-6}$	$6.84x10^{-5}$	$1.2x10^{-6}$
B	$2.83x10^{-4}$	$1.86x10^{-5}$	$3.2x10^{-3}$	$2.4x10^{-3}$
C	$6.3x10^{-4}$	$9.05x10^{-5}$	$7.2x10^{-4}$	$1.2x10^{-4}$
D	$1.86x10^{-4}$	$8.12x10^{-6}$	$1.6x10^{-3}$	$5.9x10^{-4}$
A	$8.24x10^{-4}$	$1.55x10^{-4}$	$2.4x10^{-3}$	$1.3x10^{-3}$
Mean	**$4.2x10^{-4}$**	**$5.62x10^{-5}$**	**$1.6x10^{-3}$**	**$8.9x10^{-4}$**

[a]Exposure conducted in a well-mixed batch reactor under a collimated beam.
[b]Exposure conducted in a well-mixed batch reactor with an initial chlorine concentration of 2.0 mg/L (as Cl_2).

8.8 Conclusions

Wastewater contains a large number of pathogenic microorganisms that can be transmitted by the waterborne route via swimming and drinking water. Wastewater treatment can reduce the concentrations of these microbial pathogens; however, the indicator bacteria used to traditionally judge effluent (microbial) quality do not adequately protect public health when addressing the risks associated with viruses and parasites. The risk assessment described above was based on a mean cultivable virus concentration of 75 PFU/100 L in secondary effluents. On any given day the virus concentration could be as much as ten times higher than the mean value used for these risk calculations; therefore, the risks of exposure could be an order of magnitude higher as well. Collectively, the virus inactivation estimates from the bench scale studies suggest the following:

- The risks associated with swimming in waters receiving municipal wastewater effluent range from 10^{-3} to $x10^{-6}$; risks are two to 100 times greater if the water is not disinfected, dependent on the disinfection type, intensity, and contact time, as well as the specific effluent characteristics.
- Conventional disinfection practices with chlorine may accomplish roughly 1 log_{10} virus inactivation.
- UV disinfection generally produced a better and safer effluent, in terms of effluent quality; however, the magnitude of this difference was facility-dependent and in one case, chlorination produced an equivalent effluent virus concentration.
- The wastewater industry may need to consider pathogen monitoring in effluent discharges that are impacting watersheds with significant recreational activities or that serve as a drinking water supply.

Risks of protozoan infection were not included in this risk assessment; however, UV disinfection is more effective than chlorination for inactivation of protozoa in drinking waters. No studies have been published to date regarding protozoan inactivation by UV disinfection in wastewater. Adenoviruses are the most resistant to UV disinfection and are found in high concentrations in municipal wastewater. Therefore, a transition to wastewater treatment with UV should include assessment of Adenoviruses in the future.

Also excluded from this assessment were risks associated with bacterial pathogens. Although conventional disinfection practices (UV and chlorination) generally accomplish extensive inactivation of most waterborne bacteria, it is uncommon to monitor the responses of bacterial pathogens. Moreover, the results of experiments described in Chapters 3.0 and 5.0 indicate the potential for bacteria to recover and regrow post-disinfection. It is not known how important this regrowth activity is with regard to bacterial pathogens.

Lastly, this risk assessment does not include an analysis of risk associated with chemical constituents. In the case of chlorine-based processes, disinfection has the potential to adversely affect effluent chemical composition in terms of toxicological responses. However, the inventory of DBPs that have the potential to express adverse health effects is large and highly variable among effluents. Moreover, the human health effects associated with chemical contaminants that are influenced or produced as a result of disinfection operations tend to be chronic in nature. Therefore, the development of a risk assessment for exposure to chemical constituents, including DBPs, is far more complex than the microbial risk assessment described herein.

CHAPTER 9.0

SUMMARY AND RECOMMENDATIONS

This research was designed to address two key questions:

- Should municipal wastewater effluent be disinfected prior to discharge, and
- Under circumstances where disinfection is necessary, how should it be accomplished?

More specifically, these questions were addressed from the perspective of the effects of disinfection (by chlorination/dechlorination or UV irradiation) on human health. Specific risks to human health that could be directly influenced by the application (or nonapplication) of disinfection include exposure to microbial pathogens and exposure to DBPs. In most cases, the risks posed by microbial pathogens are acute in nature, whereas those posed by DBPs tend to be chronic.

Facilities from which samples were collected were chosen to provide a broad spectrum of effluent quality, particularly as related to nitrogenous compounds. The collection and analysis of effluent samples from these facilities allowed application of a uniform set of assays across a broad range of effluent samples for assessment of the effects of common disinfectants (chlorine and UV radiation) on microbial and chemical characteristics. It is important to recognize that it is not possible in such a project to characterize all possible effluent variables. Therefore, given the case-specific characteristics of each wastewater treatment facility and their discharges (influent quality, treatment operations, volumetric flow rate, receiving water, downstream uses, etc.), it is not practical to develop a set of recommendations that will have universal applicability. However, some generalizations can be made regarding the results of this research and its implications. The purpose of this chapter is to summarize the major findings of this research and to discuss the implications of these findings.

Based on collection and analysis of samples from four municipal wastewater treatment facilities, the conclusions described below were drawn. For some assays, samples from a fifth facility were also collected and analyzed. In general, the conditions of UV irradiation and chlorination/ dechlorination employed in this research, which were designed to be representative of the conditions used in the municipal wastewater treatment facilities from which samples were collected, accomplished effective inactivation of indicator bacteria in samples evaluated immediately post-exposure. However, indicator bacteria, and more generally bacterial communities, recovered to near or above pre-exposure conditions over a period of a few days. The extent of recovery was variable, and was influenced by effluent quality and the method of disinfection.

In considering the extent of recovery within the bacterial community, several different assays were conducted. Aerobic respiration was characterized in disinfected and undisinfected effluent samples over a period of six days. In general, the application of disinfection had little, if any effect on total oxygen uptake over this period of incubation. A notable exception to this generalization appeared to exist among nitrifying bacteria. In particular, disinfection appeared to result in long-term suppression of nitrifiers; this effect was most pronounced in effluent samples that were disinfected by UV irradiation. It is not clear that suppression of nitrifier activity will have a direct bearing on human health; however, it is possible that this behavior could be important in terms of effects of wastewater effluents on the microbial ecology and chemistry of receiving waters.

Recovery in the bacterial community was also characterized by total cell counts. While disinfectants will cause a marked decline in bacterial viability immediately following disinfection, the long-term effects of these disinfectants on the bacterial population were substantially different. UV irradiation had essentially no effect on total bacterial numbers after five days. Interestingly, effluent samples that had been subjected to chlorination/dechlorination consistently showed statistically significantly higher total bacterial counts following this five-day period than either the UV irradiated or the non-disinfected samples.

FC and Ent were comparable in terms of their ability to function as bacterial indicators. Under the conditions of disinfection used in this research, both of the candidate indicators were inactivated to a similar extent. However, these two species represent a small fraction of the total bacterial population in most municipal wastewater effluents. Moreover, these organisms appear to be more sensitive to chlorination/dechlorination and UV irradiation than the general bacterial population. Therefore, their application as indicator organisms is of questionable validity.

TCB counts were examined as an alternative to conventional indicator organisms (FC and Ent). TCB were found to be roughly equivalent to FC and Ent for applications as an indicator organism; however, the TCB assay requires a longer incubation time than do assays based on FC or Ent viability. Moreover, the TCB assay provides no information regarding population composition within the bacterial community. Although the incubation period and non-specificity of the TCB assay represent limitations of the method, it is also important to recognize the benefits of this method. These include its ability to represent a broader portion of the bacterial population than conventional indicator organism assays, as well as the ability to characterize repair and recovery of the bacterial community over an extended period following disinfectant application.

The fact that bacterial assays are of questionable validity as "indicators" of possible negative human health effects is particularly important in considering the effects of wastewater disinfection on human health. In most cases, the design and operation of municipal wastewater disinfection systems is based on compliance with discharge regulations which generally limit the concentrations of viable indicator organisms (e.g., FC or Ent) and residual (halogen-based) disinfectant. However, if the objective of treatment is inactivation of organisms such as fecal coliforms and enterococci, that are known to be more sensitive to disinfectants than other microorganisms, then it is not clear that the application of disinfectants under these conditions will truly ensure adequate inactivation of the microorganisms that represent a risk to human health.

This last point is made even clearer when considering the effects of common disinfection strategies on viruses in wastewater effluents. For purposes of this research, measurements of viral responses to disinfectants were carried out using indigenous bacteriophage as viral indicators. The use of bacteriophage for this purpose was justified on the basis of their structural similarity to relevant human viruses, as well as their demonstrated resistance to common disinfectants. The results of bench-scale disinfectant exposures demonstrated that UV irradiation was, on a consistent basis, more effective than chlorination/dechlorination for inactivation of indigenous phage. In general, the conditions of batch UV irradiation used in this research were shown to be capable of inactivating most indigenous phage, although "surviving" phage were detected in several samples post-UV irradiation. In contrast, effluent samples that were subjected to chlorination/dechlorination consistently demonstrated relatively ineffective inactivation of indigenous phage. Moreover, the phage that survived chlorination/dechlorination were more diverse than those that survived UV irradiation, in terms of nucleic acid content. These observations were evident among all chlorinated effluents, but were particularly pronounced among effluent samples collected from facilities where substantial reduced N concentrations were present (i.e., facilities that did not employ nitrification). These data suggest that the application of common disinfectants to wastewater effluents, under the conditions commonly employed, may not allow for effective control of viral pathogens. This is particularly true for facilities that choose to disinfect through the application of chlorine.

A further limitation of disinfection processes is the potential for formation of DBPs. A considerable body of literature exists to describe the chemistry and by-products of chemical disinfection processes, particularly those based on chlorine. However, relatively little information was available regarding the effects of chlorination on nitrogenous compounds that are likely to exist in municipal wastewater effluents. Therefore, experiments were conducted to characterize the chemistry and toxicology of chlorination in the presence of representative organic-N compounds.

Four simple aliphatic amines were chosen as representative, or model organic-N compounds. The compounds chosen for experimentation were believed to be representative of simple organic-N compounds that are likely to be found in municipal wastewater effluent. Under the conditions of reaction that are likely to exist in most disinfection operations, chlorine was shown to interact with these compounds (predominantly) through substitution reactions. The dominant products of these reactions are mono-N-chlorinated organic amines (RHNCl). These reactions are believed to be analogous to the simple substitution reactions that yield inorganic chloramines (e.g., NH_2Cl).

Toxicity assays conducted using Microtox and by OUR tests indicated that the parent compounds (aliphatic amines) demonstrated quantifiable toxicological responses. The chlorinated forms of these compounds were found to be roughly an order of magnitude more toxic. Also, these compounds were shown to be persistent on a time scale of several days.

As a means of characterizing the effects of wastewater disinfection on human health, a risk assessment was conducted for purposes of estimating the risks of viral disease associated with disinfected or undisinfected wastewater effluent. For purposes of this risk assessment, an exposure scenario based on ingestion of a fixed volume of water, at a predefined mixing ratio, was considered for human contact by swimming. In all cases, disinfection was shown to decrease the risk of contracting disease. However, the magnitude of the risk reduction was rather modest in all cases; this was particularly true for effluent samples that were subjected to chlorination.

To place these results in the proper context, it is important to consider several aspects of the risk assessment approach in general, as well as this particular application. To conduct a risk assessment, it is necessary to assume an exposure scenario, as well as many other aspects of the system (e.g., mixing ratio, initial pathogen concentration). Obviously, it is impossible to simulate all possible exposure scenarios and all possible microbial pathogens. However, the consistent application of a reasonable set of assumptions does permit an assessment of the relative risks associated with various wastewater effluent management strategies.

Likewise, it is important to recognize that this risk assessment does not include an assessment of risks associated with exposure to other microbial pathogens (e.g., bacteria or protozoa) or chemical contaminants. Clearly, the decision to apply or not apply disinfection has the potential to affect both types of risk. In general, available information suggests that UV irradiation is likely to be more effective than chlorination for control of protozoan pathogens. In most cases, UV irradiation will also have a less pronounced effect on effluent chemistry than chlorination.

The conclusions of previous (related) research, as well as the nature of common disinfection practices need to be considered in evaluating the implications of this research. As described previously, the MWRDGC operates several large facilities for treatment of municipal wastewater, including one of the largest municipal wastewater treatment facilities in the world. Through an extensive investigation, MWRDGC was able to demonstrate that their (chlorine-based) disinfection system generally yielded decreases in receiving stream bacterial burdens only within a relatively short distance of the outfalls of their facilities (ca. <10 to 15 river miles), and had minimal effect on viruses. In terms of the health of the receiving stream, the cessation of chlorination was demonstrated to yield improvements in receiving stream ecology, both in terms of species diversity and total numbers. Moreover, the populations of pest organisms (flying insects) were observed to decrease, presumably in response to increases in fish populations. These effects were likely the result of improved toxicological conditions in the receiving stream. To reiterate, these observations were made in a nitrified effluent and in a receiving stream that at times was comprised of as much as 90% effluent. Therefore, these effects were observed in nearly undiluted effluent. The results of experiments conducted in this research, which were conducted on effluent samples from a broad range of municipal wastewater treatment facilities, are in qualitative agreement with those of the MWRDGC.

As described above, it is evident from the data provided in this report, as well as from literature, that the use of bacterial indicators such as FC and Ent is sub-optimal in some respects. The conditions that accomplish inactivation of these organisms often lead to relatively little inactivation of waterborne viruses. This argument is particularly relevant when considering chlorine-based disinfection operations and has been supported by detailed reviews of the literature pertaining to concentrations of microbial pathogens in municipal wastewater effluent and their sensitivity to common disinfectants (e.g., Sobsey, 1989).

Literature information indicates that the sensitivity of most waterborne bacteria to UV irradiation is similar to that of the common bacterial indicators (Chang et al., 1985; Sobsey, 1989). However, important exceptions to this generalization exist. For example, the spores of *Bacillus* species have been demonstrated to be substantially more resistant to UV (and most other forms of externally-applied stress) than their corresponding vegetative cells, as well as the common bacterial indicators (Nicholson and Galeano, 2003; Blatchley et al., 2004). In mild contrast with these findings, it has been reported that the resistance of waterborne bacteria to chlorination, relative to that of the common bacterial indicators, appears to be more variable than are the relative resistances to UV irradiation (Sobsey, 1989).

The protozoan parasites *C. parvum* and *G. lamblia* also represent potentially important sources of risk to human health. Extensive experimentation has demonstrated that these organisms are highly resistant to chlorine-based disinfection (e.g., Corona-Vasquez et al., 2002), whereas both organisms are highly sensitive to UV irradiation, with reported inactivation kinetics being similar to those of indicator bacteria (e.g., Qian et al., 2004).

In North America, the vast majority of municipal wastewater treatment facilities employ chlorine-based disinfection. In many northern states within the United States, disinfection is applied on a seasonal basis. For example, in Indiana most facilities disinfect from April 1 to October 31. The rationale for application of disinfection on a seasonal basis in these settings is that direct human contact is unlikely during the "cold-weather" months. Also, microbial concentrations are (typically) somewhat reduced during the nondisinfection season.

Since most municipal wastewater treatment facilities employ chlorine-based disinfection, by extension it appears that many of these facilities are probably accomplishing poor inactivation of viruses. By using bacterial indicators as the basis of design and operation, and by complying with effluent discharge regulations based on indicator bacteria and residual chlorine concentration, operators of these facilities, the regulatory community, and the general public are given the impression that disinfection systems accomplish effective control of waterborne pathogenic microorganisms. However, this research indicates that the operation of disinfection systems on this basis does not guarantee effective viral control. By comparison, UV irradiation accomplishes more effective viral control than chlorination; however, the conditions of UV irradiation commonly applied in wastewater disinfection are probably not sufficient to ensure adequate reduction of the risks associated with viral exposure when direct human contact with effluents is possible.

The effects of effluent disinfection should also be considered from the perspective of chemistry and toxicology. While UV irradiation has been demonstrated to bring about relatively little in the way of chemical or toxicological change within municipal wastewater effluents, it is well-established that chlorination does not share these traits. The results of experiments conducted in this research support previous research that indicates that chlorine-based disinfection systems generate persistent DBPs, some of which appear to express relevant toxicological characteristics.

If these arguments are correct, then it appears that disinfection systems have been relatively ineffective for control of the microbial agents within wastewaters that are most likely to cause disease. Yet, under the conditions of operation that are commonly applied in North America, the incidence of disease transmission associated with municipal wastewater effluents remains quite low. Based on these arguments, it is reasonable to question the true benefits of wastewater disinfection, as practiced today.

The first of the two central questions that formed the basis of this research was: "Should municipal wastewater effluents be disinfected prior to discharge?" No single response can appropriately answer this important question in all cases. The information presented above suggests that municipal wastewater disinfection, as commonly practiced in the United States, may not be as important in preventing communicable disease transmission as is generally assumed. It appears that control of bacterial populations is generally effective in receiving waters only within a relatively short distance from the point of discharge. Moreover, viral inactivation accomplished by most systems (particularly those that employ chlorination) is probably minimal. Therefore, in situations where direct human contact is likely, or when ingestion of indigenous organisms that are collected from the near-outfall area is likely, it appears that disinfection of municipal wastewater effluents may yield some direct benefits. However, it does not appear that the risks of disease transmission associated with these effluents is as great as it should be. In situations where direct human contact is unlikely, it is not obvious that disinfection should be employed as a default treatment process, at least not using the approaches that are common today.

With this in mind, it is also important to consider the second central question of this research: "Under circumstances where disinfection is necessary, how should it be accomplished?" In applying any disinfectant, it is critical to strike a balance between minimizing risks associated with microbial pathogens and those associated with DBPs. The data presented in this research indicate that UV irradiation and chlorination/dechlorination, when applied with the goal of complying with conventional effluent discharge regulations, are similar in terms of their ability to inactivate waterborne bacteria, although total bacterial populations generally recover to a greater extent in chlorinated effluents than in UV irradiated effluents. Perhaps more importantly, the conditions that are used to accomplish bacterial (indicator) inactivation based on chlorination/dechlorination are relatively ineffective for control of waterborne viruses, as compared with UV irradiation. Although not demonstrated in this research, it is also important to consider that the conditions of UV disinfection used in this research are likely to accomplish effective inactivation of important protozoan organisms (i.e., the (oo)cysts of *C. parvum* and *G. lamblia*). However, chlorination/dechlorination (as applied in this research and as implemented at most wastewater treatment facilities) is unlikely to be effective for inactivation of these protozoan parasites. In conjunction with these arguments, it is necessary to also consider the effects of these two common disinfection strategies on wastewater effluent chemistry and toxicity. The weight of evidence from the literature, as well as the results of experiments conducted as part of this research, indicate a clear advantage for UV irradiation in terms of chemistry and toxicology. Therefore, in circumstances where wastewater disinfection is to be applied, it appears that UV irradiation is the method of choice, based on antimicrobial efficacy, and disinfection by-products. However, decisions regarding the design, implementation, and operation of a disinfection system must be made on a site-specific basis taking into account these, as well as other relevant factors.

APPENDIX A

SUMMARY OF RANGE-FINDING EXPERIMENTS FOR DESIGN OF BENCH-SCALE DISINFECTION EXPOSURE SCENARIOS

Graphs on the following pages summarize the results of range-finding experiments that were conducted on undisinfected effluent samples from facilities studied in this research . Experiments were conducted on multiple occasions for each facility and each disinfectant.

For chlorination/dechlorination, all exposures were conducted in batch reactors with an initial chlorine concentration of 2.0 mg/L (as Cl_2). Exposures were terminated by addition of sodium thiosulfate.

UV irradiation was conducted using a collimated beam and small, shallow, well-mixed batch reactors . UV intensity was monitored using an International Light radiometer (model 1400A) with an SEL240 detector and a T2NS254 optical filter . UV dose was governed by controlling exposure time.

Viable fecal coliform concentrations were determined before and after disinfectant exposure by membrane filtration, as described in Chapter 3.0.

Figure A-1. Summary of range-finding experiments for bench-scale disinfection of effluent samples from Facility D. Top graph represents a summary of experiments based on chlorination/dechlorination. Bottom graph represents a summary of experiments based on UV_{254} irradiation.

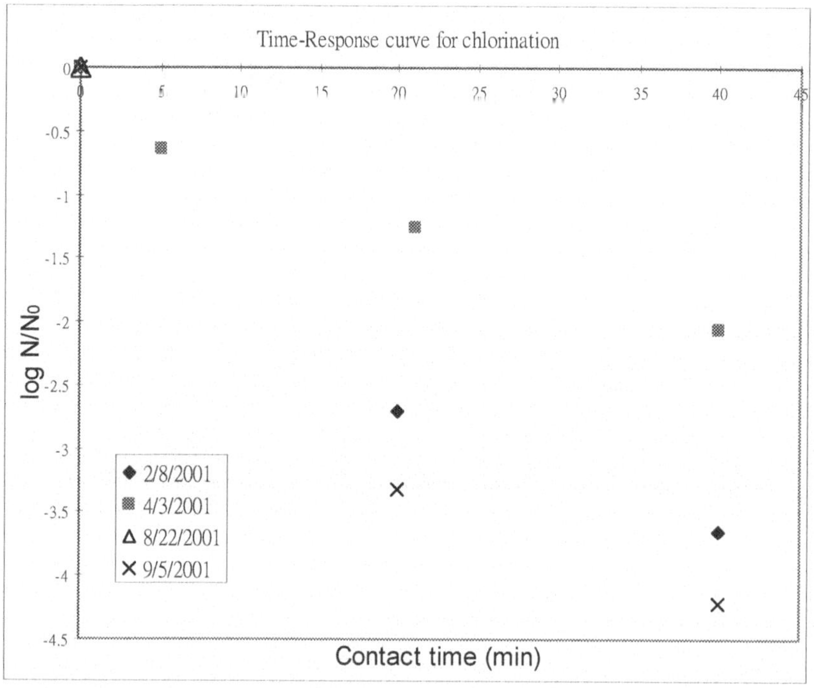

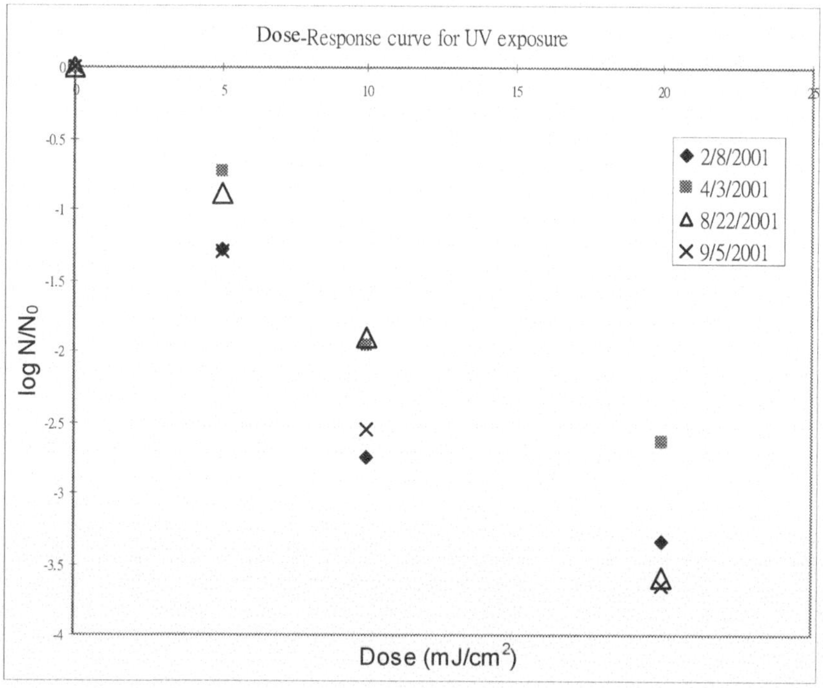

Figure A-2. Summary of range-finding experiments for bench-scale disinfection of effluent samples from Facility C. Top graph represents a summary of experiments based on chlorination/dechlorination. Bottom graph represents a summary of experiments based on UV_{254} irradiation. Arrows indicate samples that were below the limit of detection for the membrane filtration assay.

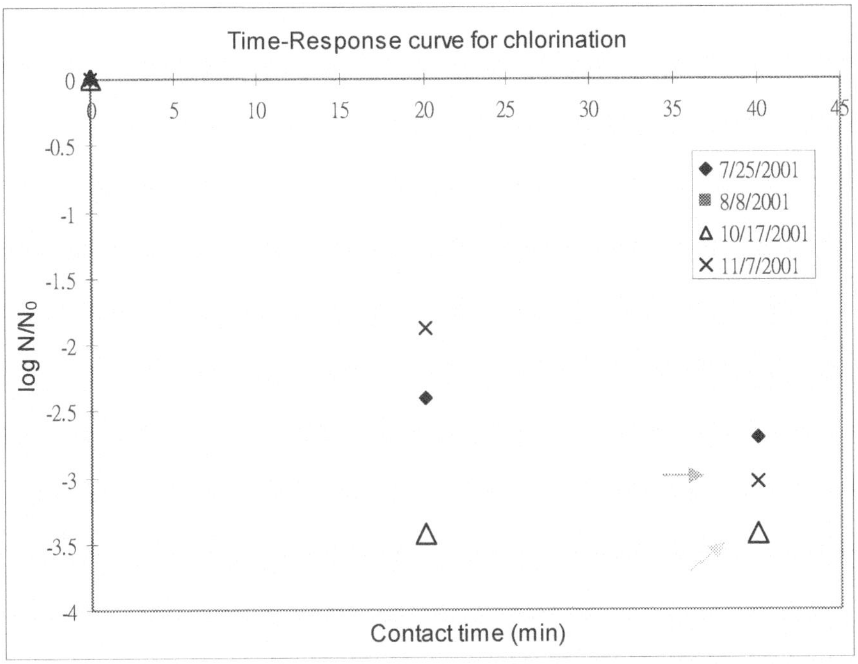

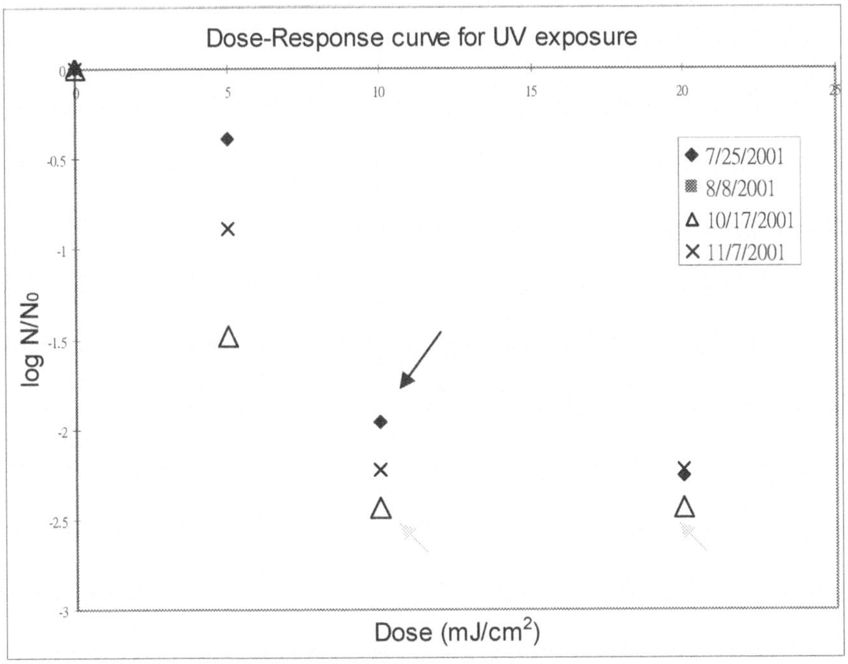

Figure A-3. Summary of range-finding experiments for bench-scale disinfection of effluent samples from Facility B. Top graph represents a summary of experiments based on chlorination/dechlorination. Bottom graph represents a summary of experiments based on UV_{254} irradiation.

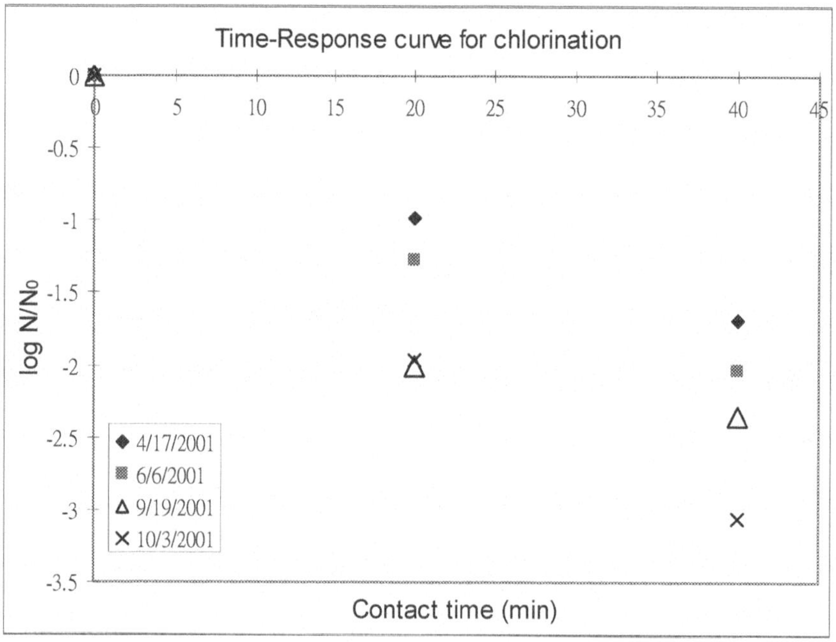

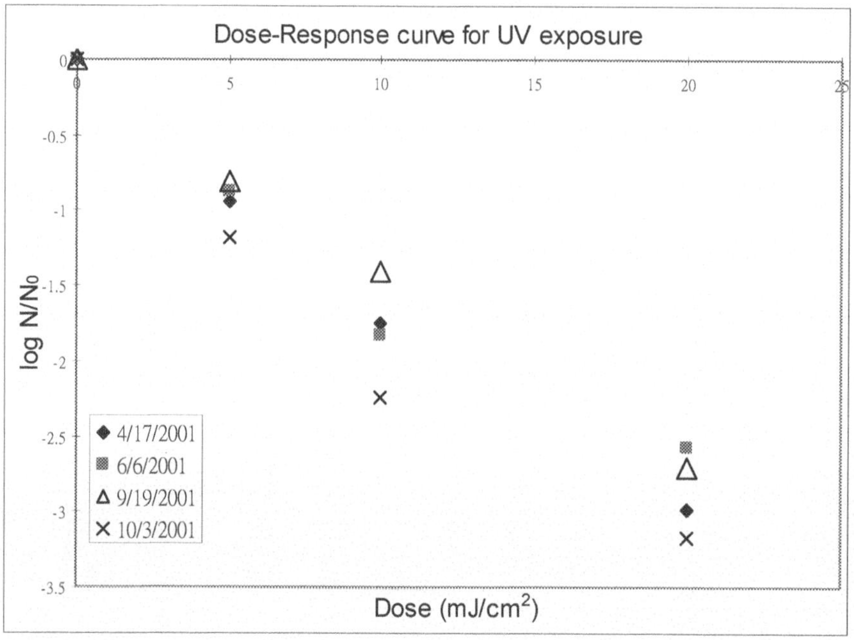

Figure A-4. Summary of range-finding experiments for bench-scale disinfection of effluent samples from Facility A. Top graph represents a summary of experiments based on chlorination/dechlorination. Bottom graph represents a summary of experiments based on UV_{254} irradiation.

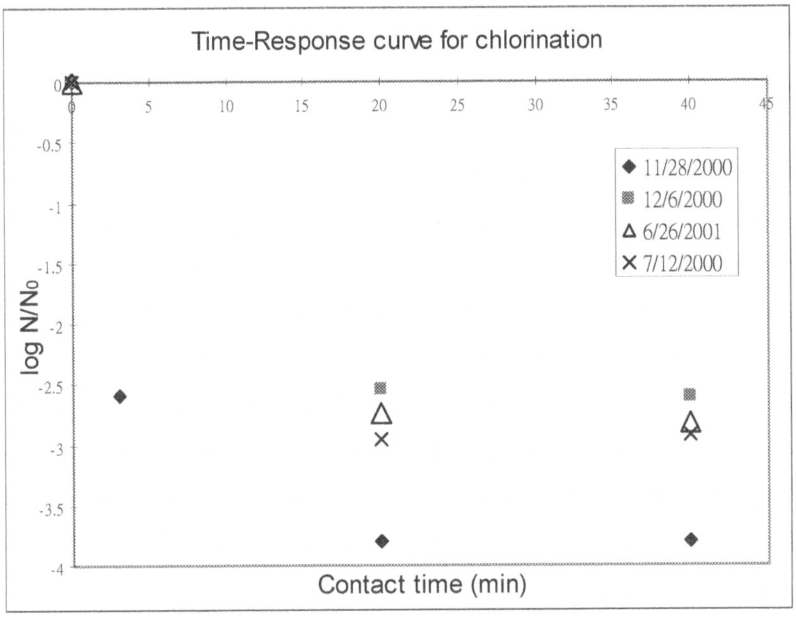

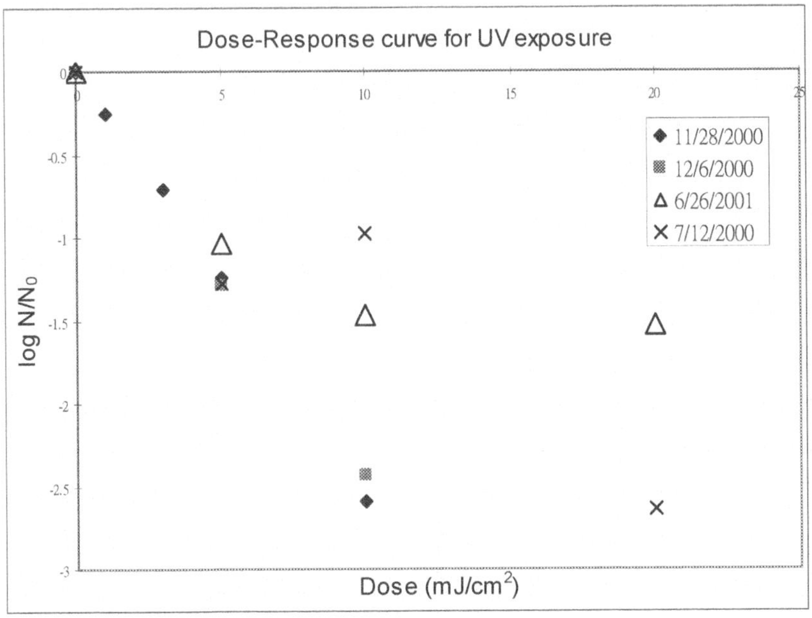

References

Acher, A.; Fischer, E.; Turnheim, R.; and Manor, Y. (1997) "Ecologically Friendly Wastewater Disinfection Techniques," *Water Research*, **31**, 6, 1398-1404.

Adams, M.H. (1959) *Bacteriophages*, Interscience, New York.

Almasi, A. and Pescod, M.B. (1996) "Pathogen Removal Mechanisms in Anoxic Wastewater Stabilization ponds," *Water Science & Technology*, **33**, 7, 133-140.

Amann, R.I.; Ludwig, W.; and Schleifer, K.H. (1995) "Phylogenetic identification and *in situ* detection of individual microbial cells without cultivation," *Microbiological Reviews*, **59,** 143-169.

American Public Health Association (1998) *Standard Methods for the Examination of Water and Wastewater* (L. Clesceri, A. Greenberg, and A. Eaton, eds.), American Public Health Association, Washington, D.C.

Anonymous (1995) "Install in Advanced Disinfection Hardware Achieves Plant Upgrade," *Water Engineering and Management,* **142**, 24.

Aoki, T. (1989) "Continuous Flow Method for Simultaneous Determination of Monochloramine, Dichloramine, and Free Chlorine: Application to a Water Purification Plant," *Environmental Science & Technology,* **23**, 46-50.

Astle, M.J. (1961) *Industrial Organic Nitrogen Compounds*, Reinhold, New York.

Atlas, R.M. and Bartha, R. (1998) *Microbial Ecology: Fundamentals and Applications*, Benjamin/Cummings, 4th Ed.

Barrette, W.; Hannum, D.; Wheeler, W.; and Hurst, J. (1989) "General mechanism for the bacterial toxicity of hypochlorous acid: abolition of ATP production," *Biochemistry*, **28**, 9172-9178.

Bauer, S. and Solyom, D. (1994) "Determination of Volatile Organic Compounds at the Parts Per Trillion Level in Complex Aqueous Matrices Using Membrane Introduction Mass Spectrometry," *Analytical Chemistry,* **66**, 24, 4422-4431.

Bauman, L.C. and Stenstrom, M.K. (1990a) "Removal of Organohalogens and Organohalogen Precursors in Reclaimed Wastewater. I," *Water Research*, **24**, 8, 949-955.

Bauman, L.C. and Stenstrom, M.K. (1990b) "Removal of Organohalogens and Organohalogen Precursors in Reclaimed Wastewater. II," *Water Research*, **24**, 8, 957-964.

Beckman Instruments (1982) "Microtox System Operating Manual Analyzer Model 2055," *Beckman Instructions 015-555879*, Microbics Operation, Carlsbad, CA.

Belhacova, L; Krysa J.; Geryk, J. and Jirkovsky, J. (1999) "Inactivation of Microorganisms in a Flow-through Photoreactor with an Immobilized TiO_2 Layer," *Journal of Chemical Technology and Biotechnology*, **74**, 149-154.

Belkin, S.; Dukan, S.; Levi, Y.; and Touati D. (1999) "Death by disinfection: Molecular approaches to understanding bacterial sensitivity and resistance to free chlorine," pp. 133-142. in E. Rosenberg (ed.), *Microbial Ecology and Infectious Disease*, American Society for Microbiology, Washington, D.C.

Berman, D.; Rice, E.; and Hoff, J. (1988) "Inactivation of particle-associated coliforms by chlorine and monochloramine," *Applied & Environmental Microbiology*, **54**, 507-512.

Bitton, G. (1994) *Wastewater Microbiology*, Wiley-Liss, New York.

Blatchley III, E.R. (1997) "Numerical modeling of UV intensity: application to collimated-beam reactors and continuous-flow systems," *Water Research*, **31,** 2205-2218.

Blatchley III, E.R. and Hunt, B.A. (1994) "Bioassay for Full-scale UV Disinfection Systems," *Water Science & Technology*, **30**, 4, 115-123.

Blatchley III, E.R.; Hunt, B.A.; Duggirala, R.; Thompson, J.E.; Zhao, J.G., Halaby, T.N.; Cowger, R.L.; Straub, C.M.; and Alleman, J.E. (1997) "Effects of Disinfectants on Wastewater Effluent Toxicity," *Water Research*, **34**, 7, 1581-1588.

Blatchley III, E.R.; Meeusen, A.; Aronson, A.I.; and Brewster, L. (2004) "Inactivation of *Bacillus* Spores by Physical Disinfectants," submitted for publication to *Water Research.*

Blum, D.J.W. and Speece, R.E. (1991) "Database of Chemical Toxicity to Environmental Bacteria and Its Use in Interspecies Comparisons and Correlations," *Research Journal of the Water Pollution Control Federation*, **63**, 3, 198-207.

Bohr, V.A.; Okumoto, D.S.; and Hanawalt, P.C. (1986) "Survival of UV Irradiated Mammalian Cells Correlated with Efficient Repair in an Essential Gene," *Proceedings of National Academic Society*, U.S.A., **83**, 3830-3833.

Bonadonna, L.; Rossella, B.; Massimo, O.; and Veschetti, E. (2002) "Occurrence of *Cryptosporidium* Oocysts in Sewage Effluents and Correlation with Microbial, Chemical, and Physical Water Variables," *Environmental Monitoring and Assessment*, **75**, 241-252.

Borrego, J.J.; Cornaxr, M.; Martinex-Manzares, E.; and Romero, P. (1990) "Coliphages as an indicator of faecal pollution in water. Their survival and productive infectivity in natural aquatic environment," *Water Research*, **24**, 111-116.

Brock, T. (1987) "The study of microorganisms *in situ*: progress and problems," pp. 1-17. in M. Fletcher (ed.) *Ecology of Microbial Communities*, Cambridge University Press, Cambridge.

Bulich, A.A. (1982) "A Practical and Reliable Method for Monitoring the Toxicity of Aquatic Samples," *Process Biochemistry*, **17**, 45-47.

Bull, R.J. (1983) In *Water Chlorination Environmental Impact and Health Effects;* (Jolley, R.L.; Brungs, W.A.; Cotruvo, J.A.; Cumming, R.B.; Mattice, J.S.; Jacobs, V.A., eds.) Ann Arbor Science; Ann Arbor, MI, 1401-1415.

Bull, R.J. and Robinson, M. (1985) In *Water Chlorination: Chemistry, Environmental Impact and Health Effects* (Jolley, R.L.; Bull, R.J.; Davis, W.P.; Katz, S.; Roberts, M.H., Jr,; Jacobs, V.A., eds.) Lewis Publishers, Chelsea, MI, **5**, 221-228.

Burton, D.T.; and Fisher, D.J. (1995) "The Residual Toxicity of Chlorine, Chlorine Dioxide and Chlorite Following Dechlorination with Sulfur IV Compounds," WREC-95-2, *WYE Research and Education Center Publications*.

Cabelli,V.J. (1983) "Water-borne viral infections," pp. 107-130. In M.Butler, A.R. Medlen, and R. Morris (eds.), *Viruses and Disinfection of Water and Wastewater*, University of Surrey Press, Guilford, England.

Calmer, J.; Neethling, J.B.; and Murbach, D. (1998) "Nitrification Effects on Chlorination and Disinfection," *Proceedings, Disnifection Specialty Conference '98*, Baltimore, MD, USA, 19-22 April, 59-70.

Campbell, H. and Striebig, B.A. (1999) "Evaluation of N-Methylpyrrolidone and Its Oxidative Products Toxicity Utilizing the Microtox Assay," *Environmental Science & Technology*, **33**, 1926-1930.

Camper, A.K. and McFeters, G.A. (1979) "Chlorine injury and the enumeration of waterborne coliform bacteria," *Applied & Environmental Microbiology*, **37**, 633-641.

Chan, Y.Y. and Killick, E.G. (1995) "The Effect of Salinity, Light and Temperature in a Disposal Environment on the Recovery of *E. coli* Following Exposure to Ultraviolet Radiation," *Water Research*, **29**, 1373-1377.

Chang, C.C. and Her, G.R. (2000) "On-line Monitoring of Trihalomethanes in Chlorinated Water by Membrane Introduction—Fast Gas Chromatography Mass-Spectrometry," *Journal of Chromatography A*, **893**, 169-175.

Chang, S.L. (1961) "Viruses, amoebas and nematodes in public water supplies," *Journal, AWWA*, **53**, 3, 188-196.

Chang, S.L.; Berg, G.; Clarke, N.A.; and Kabler, P.W. (1960) "Survival and protection against chlorination of human pathogens in free-living nematodes isolated from water supplies," *American Journal of Tropical Medicine and Hygiene*, **9**, 136-142.

Chauret, C.; Springthorpe, S.; and Sattar, S. (1999) "Fate of *Cryptosporidium* oocysts, *Giardia* cysts, and Microbial Indicators During Wastewater Treatment and Anaerobic Sludge Digestion," *Canadian Journal of Microbiology*, **45**, 257-262.

Chen, C-Y; Huang, J-B; and Chen S-D. (1997) "Assessment of the Microbial Toxicity Test and Its Application for Industrial Wastewater," *Water Science & Technology*, **36**, 12, 375-382.

Cheremisinoff, N.P. and Cheremisinoff, P.N. (1993) "Water Treatment and Waste Recovery Advanced Technology and Applications," *Prentice Hall Series in Process Pollution and Control Equipment*, NJ, 56.

Chiu, K.-P.; Lyn, D.A.; Savoye, P.; and Blatchley III, E.R. (1999a) "Integrated UV Disinfection Model Based on Particle Tracking," *Journal of Environmental Engineering, ASCE*, **125**, 1, 7-16.

Chiu, K.-P.; Lyn, D.A.; Savoye, P.; and Blatchley III, E.R. (1999b) "Effect of UV System Modifications on Disinfection Performance," *Journal of Environmental Engineering, ASCE*, **125**, 5, 459-469.

Choi, A.H.; Chang, A.T.; and Gu, M.B. (2002) "Monitoring and Classification of By-products' Toxicity using Recombinant Bioluminescent Bacterial Cell Biosensor," *Paper presented in IWA 2002 Biennial Conference*, Melbourne, Australia.

Choi, J. and Valentine, R.L. (2002) "Formation of N-nitrosodimethylamine (NDMA) from Reaction of a Monochloramine: a New Disinfection By-product," *Water Research*, **36**, 817-824.

Chong, N.M. (1994) "Biological Treatability of Amine Laden Refinery Wastewater," *Water Science & Technology*, **30,** 3, 21–28.

Choshen, E.; Johnson, J.D.; Scully, F.E., Jr.; Jersy, J.A.; Jensen, J.N.; and Jewell, J.T. (1987) "Identification of Organic N-chloramines in Wastewater," in *Water Chlorination: Chemistry, Environmental Impact and Health Effects*, **6**, 751-761.

Combs, R. and Ammerer, N. (1998) "Chlorine Removal: Ultraviolet Light Oxidation of Chlorine in Water", *Ultrapure Water*, Apr., 21-24.

Cooper, W.J. and Sorber, C.A.; Meier, E.P. (1975) "A Rapid Specific Free Available Chlorine Test with Syringaldazine" *Journal, AWWA*, **67**, 34-39.

Corona-Vasquez, B.; Samuelson, A.; Rennecker, J.L.; and Marinas, B.J. (2002) "Inactivation of *Cryptosporidium parvum* oocysts with ozone and free chlorine," *Water Research*, **36**, 4053-4063.

Cox, K.E. and Claiborne, F.B. (1949) "Effect of age and storage temperature on bacteriological water samples," *Journal, AWWA*, **41**, 948-952.

D. EPS (1990) "Guidance Document on Control of Toxicity Test Precision Using Reference Toxicants". Report EPS 1/RM/12, August 1990, Environmental Protection, Conservation and Protection, Environment Canada, p. 85.

Daniel, F.B.; Robinson, M.; Olson, G.R.; Stober, J.A. and Page, N.P. (1994) "Toxicological Studies on MX, a Disinfection By-product," *Journal, AWWA*, **86**, 3, 103-111.

Davies, C.M.; Long, J.A.; Donald, M.; and Ashbolt, N.J. (1995) "Survival of fecal microorganisms in marine and freshwater sediments," *Applied & Environmental Microbiology*, **61**, 1888-1896.

Davis, E.M. and Valentino, C.M. (1985) "Dieoff characteristics of enterococci and other indicator bacteria groups in Lake Houston," *WQTC Proceedings*, AWWA.

Davis, H.; Skrzypek, W.; and Khan, A. (1994) "Iodine Binding by Amylopectin and Stability of the Amylopectin-iodine Complex," *Journal of Polymer Science Part A – Polymer Chemistry*, **32**, 12, 2267-2274.

DeVeer, I.; Moriske, H.J.; and Ruden, H. (1994) "Photochemical Decomposition of Organic Compounds in Water After UV Irradiation: Investigation of Positive Mutagenic Effects," *Toxicology Letters*, **72**, 1/3, 113-119.

Dietz, V.J. and Roberts J.M. (2000) "National Surveillance for Infection with Cryptosporidium Parvum, 1995-1998: What Have We Learned?," *Public Health Reports*. **115**, 358-363.

Dolara, P.; Ricci, V.; Burrini, D.; and Griffini, O. (1981) "Effect of Ozonation and Chlorination on the Mutagenic Potential of Drinking Water," *Bulletin of Environmental Contamination and Toxicology*, **27,** 1-6.

Eggset, G.; Volden, G.; and Krokan, H. (1987) "Characterization of Antibodies Specific for UV Damaged DNA by ELISA," *Photochemistry and Photobiology*, **45**, 485-491.

Eggset, G.; Volden, G.; and Krokan, H. (1983) "UV-induced DNA Damage and Its Repair in Human Skin *in Vivo* Studied by Sensitive Immunohistochemical Methods," *Carcinogenesis*, **4**, 745-750.

Eilersen, A.M.; Henze, M.; and Kloft, L. (1994) "Effect of Volatile Fatty Acids and Trimethylamine on Nitrification in Activated Sludge," *Water Research*, **28**, 6, 1329-1336.

Elder, E.D.; Worley, S.D.; and William, D.E. (1987) "Bactericidal Properties of an Organic N-Chloramine Formed *In Situ*," *The Journal of Applied Bacteriology*, **62**, May, 457-464.

Elnabarawy, M.T.; Robideau, R.R.; and Beach, S.A. (1988) "Comparison of Three Rapid Toxicity Test Procedures: Microtox, Polytox and Activated Sludge Respiration Inhibition," *Toxicity Assessment*, **3**, 361 – 370.

El-Rehaili, A.M. (1995) "Response of BOD, COD and TOC of Secondary Effluents to Chlorination," *Water Research*, **29**, 6, 1571-1577.

Englebrecht, R.S. and Greenberg, E.O. (1978) "Chlorine resistant indicators," pp. 243-266. in G. Berg (ed.), *Indicators of Viruses in Water and Food*, Ann Arbor Science Publishers, Ann Arbor, MI.

Enriquez, C.E.; Christon, J.H.; and Gerba, C.P. (1995) "Survival of the Enteric Adenoviruses 40 and 41 in Tap, Sea, and Waste Water," *Water Research*, **29**(11), 2548-2553.

Esvelt, L.A.; Kaufman, W.J.; and Selleck, R.E. (1971) "Toxicity Assessment of Treated Municipal Wastewaters," *paper presented at the 44th Annual WPCF Conference*, San Francisco, CA, Oct., 4-8.

Fair, G.M.; Morris, J.C.; and Chang, S.L. (1947) "The Dynamics of Water Chlorination," *J. NEWWA*, **61**, 285.

Feng, T.H. (1966) "Behavior of Organic Chloramine in Disinfection," *J. WPCF*, **38**, 614.

Ferriol, M.; Gazet, J.; Therese, M.; and Adad, S.C. (1991) "Kinetics and Mechanisms of Chlorine Transfer from Chloramine to Amines in Aqueous Medium," *International Journal of Chemical Kinetics*, **23,** 4, 315-329.

Frost, F.J.; Craun, G.F.; and Calderon, R.L. (1996) "Waterborne Disease Surveillance," *Journal, AWWA*, **88,** 66-75.

Fry, J.C. (1990)" Direct methods and biomass estimation," pp. 41-85. in R. Grigorova and J.R. Norris (eds.) *Methods in Microbiology*, Academic Press, Cambridge, UK.

Fry, J.C. (1993) *Biological Data Analysis: A Practical Approach*, IRL Press, New York.

Gahr, A.; Weil, L.; and Niebner, R. (1995) "Polychromatic Actinometry with Filter Solutions," *Water Research,* **29,** 9, 2125-2137.

Gauthier, L.; Levi, Y.; and Jaylet, A. (1989) "Evaluation of Clastogenicity of Water Treated with Sodium Hypochlorite or Monochloramine Using a Micronucleus Test in New Larvae," *Mutagenesis*, **4**, 170-173.

Gennacarro, A.L.; Mclaughlin, M.R..; and Quintero-Betancourt, W.; Huffman, D.E.; Rose, J.B. (2003) "Infectious *Cryptosporidium parvum* Oocysts in Final Reclaimed Effluent," *Applied & Environmental Microbiology* (in press).

George, I; Crop, P.; and Servais, P. (2002) "Fecal Coliform Removal in Wastewater Treatment Plants Studied by Plate Counts and Enzymatic Methods," *Water Research*, **36**(10), 2607-2617.

Gerba C.P.; Rose, J.B.; Haas, C.N.; and Crabtree, K.D. (1996) "Waterborne Rotavirus: A Risk Assessment," *Water Research*, **30**, 2929-2940.

Gerba, C.P.; Nwachuku, N.; and Riley, K.R. (2003) "Disinfection Resistance of Waterborne Pathogens on the United States Environmental Protection Agency's Contaminant Candidate List (CCL)," *Journal of Water Supply: Research and Technology-AQUA*, **52**, 2, 81-93.

Gerba, C.P.; Rose, J.B.; and Haas, C.N. (1996a) "Sensitive Populations: Who is at the Greatest Risk?," *International Journal of Food Microbiology*, **30**, 113-123.

Godfree, A.; Kay, D.; and Wyer, M. (1997) "Faecal streptococci as indicators of faecal contamination in water," *Journal of Applied Microbiology*, **83**, S110-S119, Suppl. S.

Goldschmidt, S.; Wiberg, E.; Nagel, F.; and Martin, K. (1927) *Justus Liebigs Annalen der Chemie*, 456, 1-38.

Gowda, N.M.M.; Trieff, N.M.; and Stanton, G.J. (1986) "Kinetics of Inactivation of Adenovirus in Water by Chloramines-T," *Water Research*, **20**, 817-823.

Grabow, W.O.K.; Coubrough, P.; Nupen, E.M.; and Bateman, B.W. (1984) "Evaluation of coliphages as indicators of the virological quality of sewage-polluted water," *Water SA* **10**, 7-14.

Grabow, W.O.K.; Holtzhausen,C.S.; and DeVilliers, J.C. (1993) "Research on bacteriophages as indicators of water quality," WRC Report No 321/1/93, Water Research Commission Pretoria. 147 pp.

Grabow, W.O.K.; Neubrech, T.E.; Holtzhausen, C.S.; and Jofre, J. (1995) *Bacteriodes fragilis* and *Escherichia coli* bacteriophages: excretion by humans and animals. *Water Science &.* Technology, **31**, 223-230.

Gronberg, L. and Jonsson, J.A. (1992) "Measurement of Aliphatic Amines in Ambient Air and Rainwater," *Chemosphere,* **24**, 10, 1533–1540.

Guerrero, M.A. and Jones, R.D. (1996) "Photoinhibition of Marine Nitrifying Bacteria. I. Wavelength-dependent Response," *Marine Ecology Progress Series*, **141**, 183-192.

Guter, K.J.; Cooper, W.J.; and Sorber, C.A. (1974) "Evaluation of Existing Field Test Kits for Determining Free Chlorine Residuals in Aqueous Solutions," *Journal, AWWA*, **66**, 38-48.

Haas, C.H..; Rose, J.B.; and Gerba, C.P. (eds) (1999) *Quantitative Microbial Risk Assessment*, John Wiley and Sons, New York.

Haas, C.N.; Crockett, C.S.; Rose, J.B.; Gerba, C.P., and Fazil, A.M. (1996) "Assessing the Risk Posed By Oocysts in Drinking Water," *Journal, AWWA*, **88**, 9, 131-136.

Haas, C.N.; Rose, J.B.; Gerba, C.P.; and Regli, S. (1993) "Risk Assessment of Virus in Drinking Water," *Risk Analysis*, **13**, 5, 545-552.

Haas, C.N.; Thayyat-Madabusi, A.; Rose, J.B.; and Gerba, C.P. (2000) "Development of a Dose-Response Relationship for *Escherichia coli* 0157:H7," *International Journal of Food Microbiology*, **1**, 1-7.

Haas, C.N.; Hornberger, J.C.; Anmangadla, U.; Heath, M.; and Jacangelo, J. (1993) Experimental Methodologies for the Determination of Disinfection Effectiveness. *AWWA Research Foundation.*

Hao, O.J.; Shin, C-J; Lin, C-F; Jeng, F-T; and Chen Z-C. (1996) "Use of Microtox Tests for Screening Industrial Wastewater Toxicity," *Water Science & Technology*, **34**, 10, 43-50.

Harp, D.L. (2002) *Current Technology of Chlorine Analysis for Water and Wastewater*, Technical Information Series—Booklet No. 17, Hach Company, Loveland, CO.

Harris, G.D.; Adams, D.; Moore, W.M.; and Sorensen, D.L. (1987) "Potassium Ferrioxalate as Chemical Actinometer in Ultraviolet Reactors," *Journal of Environmental Engineering, ASCE*, **113**, 612.

Haugland, R. (1996) *Handbook of Fluorescent Probes and Research Chemicals*, Molecular Probes, Inc., Eugene, OR.

Havalaar, A.H.; Furuse, K.; and Hogeboom, W.M. (1986) "Bacteriophages and indicator bacteria in human and animal faeces," *Journal of Applied Bacteriology*, **60**, 3, 255-262.

Havalaar, A.H. and Hogeboom, W.M. (1983) "Factors Affecting the Enumeration of Coliphages in Sewage and Sewage Polluted Waters," *Antonie van Leeuwenhoek Journal of Microbiology*, **49**, 387-397.

Havalaar, A.H.; Pot-Hogeboom, W.M.; Furuse, K.; and Hormann, P.R. (1990) "F-specific RNA bacteriophages and sensitive strains in faeces and wastewater of human and animal origin. *Journal of Applied Bacteriology*, **69**, 30-37.

Heath, M.(1995) "Comparative Costs of Chlorination/Dechlorination and UV Radiation," *WEFTEC '95, 68th Annual Conference and Exposition*, **5**, 227.

Helz, G.R. andNweke, A.C. (1995) "Incompleteness of Wastewater Dechlorination," *Environmental Science & Technology*, **29**, 1018-1022.

Hoben, H.J. and Somasegaran, P. (1982) "Comparison of the pour, spread, and drop plate methods for enumeration of *Rhizobium* spp. in inoculants made from presterilized peat," *Applied & Environmental Microbiology*, **44**, 1246-1247.

Hofman, A.W. and Frankland, E. (1859) *Report on the Deodorization of Sewage*, Report to the Metropolitan Board of Works, London, Aug. 12, 1859.

Holm-Hansen, O.; Lubin, D.; and Helbling, E.W. (1993) "Ultraviolet Radiation and Its Effects on Organisms in Aquatic Environments," *Environmental UV Photobiology*, edited by Antony R. Young *et al*., Plenum Press, 379-425. New York.

Huang, C.T.; Yu, F.P.; McFeters, G.A.; and Stewart, P.S. (1995) "Nonuniform spatial patterns of respiratory activity within biofilms during disinfection," *Applied & Environmental Microbiology*, **61**, 2252-2256.

Hureiki, L.; Croue, J.P.; and Legube, B. (1994) "Chlorination Studies of Free and Combined Amino Acids," *Water Research*, **28,** 12, 2521-2531.

Hwang, Y.W.; Matsuo, T.; Hanaki, K.; Suzuk, N. (1995) "Identification and Quantification of Sulfur and Nitrogen Containing Odorous Compounds in Wastewater," *Water Research*, **29,** 2, 711-718.

IAWPRC Study Group on Health Related Water Microbiology (1991) "Bacteriophages as model viruses in water quality control," *Water Research*, **25**, 529-545.

International Standards Organization (1984) "Water Quality-activated Sludge Oxygen Consumption Inhibition Test," DIS 8192, Geneva.

Isaac, R.A.; Morris, J.C.; Jolley, R.L; Brungs, W.A. and Cumming, R.B., Eds. (1980) *Water Chlorination: Environmental Impacts and Health Effects*, Ann Arbor Science, Ann Arbor, MI, **3**, Chapter 17.

Isaac, R.A. and Morris, J.C. (1983) "Transfer of Active Chlorine from Chloramine to Nitrogenous Organic Compounds - 1. Kinetics," *Environmental Science & Technology*, **17**, 738.

Jacangelo, J.G.; Olivieri, V.P.; and Kawata, K. (1987) "Mechanism of inactivation of microorganisms by combined chlorine," American Water Works Association Publishers, Denver.

Jacangelo, J.G.; Loughran, P.; Petrik, B.; Simpson, D.; and McIlroy, C. (2003) "Removal of Enteric Viruses and Selected Microbial Indicators by UV Irradiation of Secondary Effluent," *Water Science & Technology*, **47**(9), 193-198.

Jensen, J.N. and Johnson, J.D. (1990a) "Interferences by monochloramine and organic chloramines in free available chlorine methods. 1. Amperometric titration," *Environmental Science & Technology,* **24**, 7, 981-984.

Jensen, J.N. and Johnson, J.D. (1990b) "Interferences by monochloramine and organic chloramines in free available chlorine methods. 2. N,N-diethyl-p-phenylenediamine," *Environmental Science & Technology*, **24**, 7, 985-990.

Jensen, J.S. (1997) *Chemical Studies to Understand the Dechlorination Process Used at Wastewater Treatment Plants*, Ph. D. dissertation, University of Maryland, p145.

Jensen, J.S. and Helz, G.R. (1998a) "Dechlorination Kinetics at Alkaline pH of N-chloropiperidine, a Genotoxin in Chlorinated Municipal Wastewater," *Water Research*, **32,** 9, 2615-2620.

Jensen, J.S. and Helz, G.R. (1998b) "Rates of Reduction of N-chlorinated Peptides by Sulfite: Relevance to Incomplete Dechlorination of Wastewaters," *Environmental Science & Technology*, **32,** 4, 516-522.

Jopling, W. (1975) paper presented by the California State Department of Public Health at a joint meeting of Arizona, California, Nevada, and EPA regarding waste discharge requirements and water quality objectives for the Colorado River, Las Vegas, NV, Oct. 28, 1975.

Kaiser, K.L.E. and Ribo, J.M. (1988) "*Photobacterium Phosphoreum* Toxicity Bioassay, II. Toxicity Data Compilation," *Toxicological Assessment*, **3**, 195-237.

Kaprelyants, A. and Kell, D. (1992) "Rapid assessment of bacterial viability and vitality by rhodamine 123 and flow cytometry," *Journal of Applied Bacteriology* **72**, 410-422.

Kashimada, K.; Kamiko, N.; Yamamoto, K.; and Ohgaki, S. (1996) "Assessment of Photoreactivation Following Ultraviolet Light Disinfection," *Water Science & Technology*, **33,** 10-11, 261-269.

Keswick, B.H. (1982) "Monitoring disinfection of enteric viruses with baceriophages," *Viruses and Disinfection of Water and Wastewater Proceeding of the International Symposium*, University of Surry, Guildford.

Kijak, P.J. and Helz, G.R. (1988) "Fate of Sulfur (IV) Dechlorinating Agents in Natural Waters: Effects of Suspended Sediments," *Environmental Science & Technology*, **22,** 10, 1171-1177.

King, E.F. and Painter, H.A. (1985) "The Effects of Acclimatization on the Toxicity of Chemicals to Activated Sludge Microorganisms," *Environment Pollution Series A*, ***1***, 39, 267-280.

Kinniment, S.L. and Wimpenny, J.W. (1992) "Measurements of the distribution of adenylate concentrations and adenylate energy charge across *Pseudomonas aeruginosa*," *Applied & Environmental Microbiology*, **58**, 1629-1635.

Koivunen, J.; Siitonen, A.; and Heinonen-Tanski, H. (2003) "Elimination of Enteric Bacteria in Biological-Chemical Wastewater Treatment and Tertiary Filtration Units," *Water Research*, **37**, 690-698.

Kotiaho, T.; Hayward, M.J.; and Cooks, R.G. (1991) "Direct Determination of Chlorination Products of Organic Amines Using Membrane Introduction Mass Spectrometry," *Analytical Chemistry*, **63**, 17, 1794–1801.

Kotiaho, T.; Lauritsen, F.R.; Choudhury, T.K.; Cooks, R.G.; and Tsao, G.T. (1991) "Membrane Introduction Mass Spectrometry," *Analytical Chemistry*, **63**, 18, 875A-883A.

Kott,Y.; Ben-Ari, H.; Vinokur, L. (1978) "Coliphage survival as viral indicators in various wastewater quality effluents," *Progress in Water Technology*, **10**, 337-346.

Kott, Y.; Roze, N.; Sperber, S.; Betzer, N. (1974) "Bacteriophages as viral pollution indicators," *Water Research*, **8**, 165-171.

Krasner, S.W.; McGuire, M.J.; Jacangelo, J.G.; Patania, N.L.; Reagan, K.M.; Aieta, E.M. (1989) "Occurrence of Disinfection By-products in U.S. Drinking Water," *Journal AWWA*, **81**, 8, 41-53.

Larson, R.J. and Schaeffer, S.L. (1982) "A Rapid Method for Determining the Toxicity of Chemicals to Activated Sludge," *Water Research*, **16**, 675-680.

Lazarova, V.; Savoye, P.; Janex, M.L.; Blatchley III, E.R.; and Pommepuy, M. (1999) "Advanced Wastewater Disinfection Technologies: State of The Art and Perspectives," *Water Science & Technology*, **40**, 4, 203-213.

Lazarova, V.; Janex, M.L.; Fiksdal, L.; Oberg, C.; Barcina, I.; and Pommepuy, M. (1998) "Advanced Wastewater Disinfection Technologies: Short and Long Term Efficiency," *Water Science & Technology*, **38**, 12, 109-117.

LeChevalier, M.W.; Babcock, T.M.; and Lee, R.G. (1987) "Examination and Characterization of Distribution System Biofilms," *Applied & Environmental Microbiology*, **53,** 12**,** 2714-2724.

LeChevallier, M.W.; Cawthon, C.D.; and Lee, R.G. (1988) "Factors promoting survival of bacteria in chlorinated water supplies," *Applied & Environmental Microbiology*, **54**, 649-654.

LeChevallier, M.W.; Evans, T.M.; and Seidler, R.J. (1981) "Effect of turbidity on chlorination efficiency and bacterial persistence in drinking water," *Applied & Environmental Microbiology*, **42**, 159-167.

Lee, S.H.; Levy, D.A.; Craun, G.F.; Beach, M.J.; Calderon, R.L. (2002) "Surveillance for Waterborne-Disease Outbreaks—United States, 1999-2000," *MMWR*, **51**(SS-8), 1-47.

Lenwood, Jr., W.H.; Dennis, T.B.; William, C.G.; and Stuart, L.M. (1981) "Effects of Dechlorination on Early Life Stages of Striped Bass (*Morone saxatilis*)," *Environmental Science & Technology*, **15,** 5, 573-578.

Levenspiel, O. (1972) *Chemical Reaction Engineering*, 2nd edition, 326-348, John Wiley & Sons, New York.

Lind, C. (1995) "Wastewater Dechlorination Options," *Public Works*, Sep., 69-70.

Linden, K.G. and Darby, J. (1997) "Estimating Effective Germicidal Dose from Medium Pressure UV Lamps," *Journal of Environmental Engineering, ASCE*, **123**, 1142.

Linden, K.G.; Shin, G.; Faubert, G.; Cairns, W.; and Sobsey, M.D. (2002) "UV Disinfection of *Giardia lamblia* Cysts in Water," *Environmental Science & Technology*, **36**(11), 2519-2522.

Lisle, J.; Pyle, B.; and McFeters, G. (1999) "The use of multiple indices of physiological activity to assess viability in chlorine disinfected *Escherichia coli* O157:H7," *Letters in Applied Microbiology*, **29**, 42-47.

Lisle, J.; Broadaway, S.; Prescott, A.; Pyle, B.; Fricker, C.; and McFeters, G. (1998) "Effects of starvation on physiological activity and chlorine disinfection resistance in *Escherichia coli* O157:H7," *Applied & Environmental Microbiology*, **64**, 4658-4662.

Liu, D. (1981) "A Rapid Biochemical Test for Measuring Chemical Toxicity," *Bulletin of Environmental Contamination and Toxicology*, **26**, 145-149.

Liu, D. and Dutka, B.J. eds. (1984) *Toxicity Screening Procedures Using Bacterial Systems*, Marcel Dekker, New York.

Lu, S.K. and Lederberg, J. (1976) "Chloramine Mutagenesis in *Bacillus subtilis*," *Science*, **192**, 1141-1143.

Lukasewycz, M.T.; Bieringer, C.M.; Liukkonen, R.J.; Fitzsimmons, M.E.; Corcoran, H.F.; Lin, S.; and Carlson, R.M. (1989) "Analysis of Inorganic and Organic Chloramines: Derivatization with 2-mercaptobenzothiazole," *Environmental Science & Technology*, **23**, 2, 196-199.

Ma, Y-S and Lin, J-G. (1998) "Effect of Pre-sonication on Removal of Organic Matters Resulting from Chlorinated Humic Acids," *Water Science & Technology,* **38,** 6, 5, 253-260.

MacCrehan, W.A.; Jensen, J.S.; and Helz, G.R. (1998) "Detection of Sewage Organic Chlorination Products that are Resistant to Dechlorination with Sulfite," *Environmental Science & Technology,* **32,** 22, 3640-3645.

MacKenzie, W.R.; Hoxie, N.J.; Proctor, M.E.; Gradus, M.S.; Blair, K.A.; Peterson, D.E., Kazmierczak, J.J.; Addiss, D.G.; Fox, K.R.; Rose, J.B. (1994) "A Massive Outbreak in Milwaukee of *Cryptosporidium* Infection Transmitted Through The Public Water Supply," *New England Journal of Medicine*, **331**, 161-167.

Mahin, T. and Pancorbo, O. (1999) "Waterborne Pathogens," *Water Environmental & Technology*, **11**, 4, 51-55.

Margerum, D.W.; Gray, E.T. Jr.; and Huffman, R.P. "Chlorination and the Formation of N-chloro Compounds in Water Treatment," *Organometals and Organometalloids, Occurrence and Fate in the Environment*, ACS Symposium Series, **82**, Washington, D.C., 278-291, 1978.

Matin, A. and Harakeh, S. (1990) "Effect of starvation on bacterial resistance to disinfectants," pp. 88-103. in G. McFeters (ed.) *Drinking Water Microbiology*, Springer-Verlag, New York.

Matsunaga, I.; Tomoda, R.; Nakajima, T.; Nakamura N.; and Komine T., (1988) "Continuous Sterilization System that Uses Photosemiconductor Powders," *Applied & Environmental Microbiology*, **54**, 1330-1333.

Mayne, L.V. and Lehmann, A.R. (1982) "Failure of RNA Synthesis to Recover After UV Irradiation: an Early Defect in Cells from Individuals with Cockayne's Syndrome and Xeroderma Pigmentosum," *Cancer Research*, **42**, 1473-1478.

McDaniels, A.E. and Bordner, R.H. (1983) "Effects of holding time and temperature on coliform numbers in drinking water," *Journal, AWWA*, **75**, 458-463.

McFeters, G. (1990) "Enumeration, occurrence, and significance of injured indicator bacteria in drinking water," pp. 478-492. in G. McFeters (ed.) *Drinking Water Microbiology*, Springer-Verlag, New York.

McFeters, G. (1997) "Effects of aquatic environmental stress on enteric bacterial pathogens and coliforms," pp. 235-242. in D. Kay (ed.) *Coliforms and E. coli: Problem or Solution?* The Royal Society of Chemistry, Cambridge, UK.

McFeters, G.A. and Camper, A.K. (1983) "Enumeration of indicator bacteria exposed to chlorine," *Advances in Applied Microbiology*, **29**, 177-193.

McFeters, G.A.; Kippin, J.S.; and LeChevallier, M.W. (1986) "Injured coliforms in drinking water," *Applied & Environmental Microbiology*, **51**, 1-5.

McFeters, G.A.; Broadway, S.C.; Pyle, B.H.; and Egozy, Y. (1993) "Distribution of bacteria within operating laboratory water purification systems," *Applied & Environmental Microbiology*, **59**, 1410-1415.

McFeters, G.; Pyle, B.; Lisle, J.; and Broadaway, S. (1999) "Rapid direct methods for enumeration of specific, active bacteria in water and biofilms," *Journal of Applied Microbiology*, **85**, 193S-200S.

McFeters, G.; Yu, F.; Pyle, B.; and Stewart, P. (1995) "Physiological assessment of bacteria using fluorochromes," *Journal of Microbial Methods*, **21**, 1-13.

Mellon, I.; Bohr, V.A.; Smith, C.A.; and Hanawalt, P.C. (1986) "Preferential DNA Repair of an Actively Transcribed Gene in Human Cells," *Proceeding of National Academy of Science*, U.S.A., **83**, 8878-8882.

Mena, K.D.; Gerba, C.P.; Haas, C.N.; and Rose, J.B. (2003) "Risk Assessment of Waterborne Coxsackievirus," *Journal, AWWA*, **95**(7), 122-131.

Meng, Q.S. and Gerba, C.P. (1996) "Comparative inactivation of enteric adenoviruses, poliovirus and coliphages by ultraviolet irradiation," *Water Research*, **30**, 2665-2668.

Meyland, S.J. and Machlin, A. (2001) *The Plain Language Guide to Sanitary Sewer Overflows: A Brief Introduction to the Basics of SSOs and What You Can Do To Eliminate Them.* Citizens Environmental Research Institute, Farmingdale, NY.

Microbics Corporation (1992) *Microtox Manual*, Microbics Corporation, Carlsbad, CA..

Mink, F.L.; Coleman, W.E.; Munch, J.W.; Kaylor, W.H.; and Ringland, H.P. (1983) *Bulletin of Environmental Contamination and Toxicology*, **30**, 394-399.

Mitch, W.A. and Sedlak, D.L. (2002) "Formation of N-Nitrosodimethylamine (NDMA) from Dimethylamine during Chlorination," *Environmental Science & Technology*, **36**, 4, 588-595.

Moan, J. (1989) "Effects of UV Radiation on Cells," *Journal of Photochemistry and Photobiology B*, **4**, 21-34.

Mofidi, A.A.; Baribeau, H.; Rochelle, P.A.; De Leon, R.; Coffey, B.M.; and Green, J.F. (2001) "Disinfection of *Cryptosporidium parvum* with Polychromatic UV Light", *Journal, AWWA*, **93**, 6, 95-109.

Monarca, S.; Feretti, D.; Collivignarelli, C.; Guzzella, L.; Zerbini, I.; Bertanza, G.; and Pedrazzani, R. (2000) "The Influence of Different Disinfectants on Mutagenicity and Toxicity of Urban Wastewater," *Water Research*, **34,** 17, 4261-4269.

Moore, L.W. and Baker, W.L. (1996) "Solving a Complex Chlorination," *Water Environment & Technology*, December, 41–46.

Nacci, D.; Jackim, E.; and Walsh, R. (1986) "Comparative Evaluation of Three Rapid Marine Toxicity Tests: Sea Urchin Early Embryo Growth Test, Sea Urchin Sperm Cell Toxicity Test and Microtox," *Environmental Toxicology & Chemistry*, **5**, 521-525.

National Research Council (1998) *Issues in Potable Reuse: The Viability of Augmenting Drinking Water Supplies with Reclaimed Water*. National Academy Press, Washington, D.C.

Nebe-von Caron, G.; Stephens, P.; and Badley, R. (1998) "Assessment of bacterial viability status by flow cytometry and single cell sorting," *Journal of Applied Microbiology*, **84**, 988-998.

Ng, W.J.; Ong, S.L.; and Hu, J.Y. (2001) "The Effects of Water Reclamation Technologies on Biological Stability of Industrial Water," *Water Science & Technology*, **43,** 10, 327-334.

Nicholson, W.L. and B. Galeano. (2003) "UV resistance of *Bacillus anthracis* spores revisited: Validation of *Bacillus subtilis* spores as UV surrogates for spores of *B. anthracis* Sterne," *Applied &. Environmental Microbiology*, **69**, 1327-1330

Nouraldeen, A.M. and Ahmed. A.E. (1996) "Studies on the Mechanisms of Haloacentronitrile-Induced Genotoxicity IV: *In vitro* Interaction of Haloacetonitriles with DNA," *Toxicology in Vitro*, **10,** 1, 17-26.

Novotny, C.P. and Lavin, K. (1971) "Some effects of temperature on the growth of F pili," *Journal of Bacteriology*, **9**, 671-682.

Nowell, L.H. and Hoigne, J. (1992a) "Photolysis of Aqueous Chlorine at Sunlight and Ultraviolet Wavelengths - I. Degradation Rates," *Water Research,* **26,** 5, 593-598.

Nowell, L.H. and Hoigne, J. (1992b) "Photolysis of Aqueous Chlorine at Sunlight and Ultraviolet Wavelengths - II. Hydroxyl Radical Production," *Water Research,* **26,** 5, 599-605.

Nweke A. and Scully F.E., Jr. (1989) "Stable N-Chloroaldimines and Other Products of the Chlorination of Isoleucine in Model Solutions and in a Wastewater," *Environmental Science & Technology*, **23**, 989-994.

Nweke, A.C. and Helz, G.R. (1990) "A Method for Determining the Completeness of Dechlorination with Sulfite," *in Water Chlorination: Chemistry. Environment Impact and Health Effects*, Jolley, R. L., et al., Eds.; Lewis Publishers: Chelsea, MI; Vol. 6, 741-750.

Oliver, J.D. (1992) "Formation of Viable but Non-culturable Cells. In Starvation in Bacteria," Kjelleberg Ed., Plenum Press, **4**, 239-272.

Olson, B.H. and Stewart, M. (1987) "Factors that change bacterial resistance to disinfection," pp. 885-904. in R. L. Jolley (ed.) *Water Chlorination: Chemistry, Environmental Impact and Health Effects*, Lewis Publishers, Chelsea, MI.

Olson, B.H. (1982) EPA-600/52-82-072.

Ono, Y.; Somiya, I.; and Kawaguchi, T. (1996) "Genotoxicity of Substances in the Nightsoil and its Biologically Treated Water," *Water Research*, **30**, 3, 569-577.

Ono, Y.; Somiya, I.; and Oda, Y. (2000) "Identification of a Carcinogenic Heterocyclic Amine in River Water," *Water Research*, **34**, 3, 890-894.

Ono, Y.; Somiya, I.; Kawamura, M.; and Uenish, K. (1992) "Genotoxicity of Organic Substances in Municipal Sewage and its Ozonated Products," *Water Science & Technology*, **25,** 11, 285-291.

Oppenheimer, J.A.; Jacangelo, J.G.; Laine, J-M.; and Hoagland, J.E. (1997) "Testing the Equivalency of Ultraviolet Light and Chlorine for Disinfection of Wastewater to Reclamation Standards," *Water Environment Research,* **69**, 1, 14-24.

Osgood, C. and Sterling, D. (1991) "Dichloroactonitrile, a By-product of Water Chlorination, Induces Aneuploidy in Drosophila," *Mutation Research*, **261**, 2, 85-91.

Oyamada, M.; Igarashi, K.; and Kuwatsuka, S. (1980) "Degradation of the Herbicide Naproanilide, 1-(2-naphthoxy) Propionanilide**,** in Flooded Soils Under Oxidative and Reductive Conditions," *Nippon Noyaku Gakkaishi.* [Tokyo, Pesticide Science Society of Japan], **5**, 4, 495-501.

Pace, N.R. (1996) "New Perspectives on the Natural Microbial World: Molecular Microbial Ecology," *ASM News*, ***62***,463-470.

Payment, P.; Plante, R.; and Cejka, P. (2001) "Removal of Indicator Bacteria, Human Enteric Viruses, *Giardia* cysts, and *Cryptosporidium* oocysts at a Large Wastewater Primary Treatment Facility," *Canadian Journal of Microbiology*, **47**(3), 188-193.

Pedersen, E.J.; Urbansky, E.T.; Marinas, B.J.; and Margerum D.W. (1999) "Formation of Cyanogen Chloride from the Reaction of Monochloramine with Formaldehyde," *Environmental Science & Technology*, **33**, 4239-4249.

Pett, B. (1993) "Cryptosporidiosis Outbreak from an Operation Point of View," *Proceedings, 1993 AWWA WQTC, Miami.*

Pinto, B.; Pierotti, R.; Canale, G.; and Reali, D. (1999) "Characterization of fecal streptococci as indicators of faecal pollution and distribution in the environment," *Letters in Applied Microbiology*, **29**, 258-263.

Porter, K.G.; Feig, Y.S. (1980) "The use of DAPI for identifying and counting aquatic microflora," *Limnology & Oceanography*, **25**, 943-948.

Puig, A.; Jofre, J.; Araujo, R. (1998) "Use of oxyrase enzyme (Oxyrase) for the detection of bacteriophages of *Bacteroides fragilis* in aerobic incubating conditions," *Journal of Microbial Methods*, **31**, 205-207.

Pyle, B.; Broadaway, S.; and McFeters, G. (1995) "A rapid, direct method for enumerating respiring enterohemorrhagic *Escherichia coli* O157:H7 in water," *Applied & Environmental Microbiology*, **61**, 2614-2619.

Qian, S.S.; Donnelly, M.; Schmelling, D.C.; Messner, M.; Linden, K.G.; and Cotton, C. (2004) "Ultraviolet light inactivation of protozoa in drinking water: A Bayesian meta-analysis," *Water Research*, **38**, 317-326.

Qureshi, A.A.; Flood, K.W.; Thompson, S.R.; Janhurst, S.M.; Inniss, C.S. and Rokosh, D.A. (1982) "Comparison of A Luminescent Bacterial Test with Other Bioassays for Determining Toxicity of Pure Compounds and Complex Effluents," *Aquatic Toxicology and Harzard Assessment: Fifth Conference* (Pearson, J.G.; Foster, R.B.; Bishop, W.E., eds.) ASTM STP 766, American Society for Testing and Materials, Philadelphia, p 179-195.

Rahn, R.O. (1997) "Potassium Iodide as a Chemical Actinometer for 254 nm Radiation: Use of Iodate as an Electron Scavenger," *Photochemistry and Photobiology*, **66,** 4, 450-455.

Rahn, R.O. (1993) "Use of Potassium Iodide as a Chemical Actinometer," *Photochemistry and Photobiology*, **58,** 6, 874-880.

Ram, N.M. and Malley, J.P. Jr. (1984) "Chlorine Residual Monitoring in the Presence of N-organic Compounds," *Journal, AWWA*, **76**, 9, 74-81.

Ram, N.M. and Morris, J.C. (1980) "Environmental Significance of Nitrogenous Organic Compounds in aquatic Sources," *Environment International,* **4**, 397-405.

Ram, N.M. and Morris, J.C. (1982) "Selective Passage of Hydrophilic Nitrogenous Organic Materials Through Macroreticular Resins," *Environmental Science & Technology*, **16**, 3, 170-174.

Reasoner, D. and Geldreich, E. (1985) "A new medium for the enumeration and subculture of bacteria from potable water," *Applied & Environmental Microbiology*, **49**, 1-7.

Rebhun, M.; Heller-Grossman, L.; and Manka, J. (1997) "Formation of Disinfection Byproducts During Chlorination of Secondary Effluent and Renovated Water," *Water Environment Research*, **69,** 6, 1154-1161.

Rees, G.; Pond, K.; Johal, K.; Pedley, S.; and Rickards, A. (1998) "Microbiological analysis of selected coastal bathing waters in the U.K., Greece, Italy, and Spain," *Water Research*, **32**, 2335-2340.

Regli, S.; Rose, J.B.; Haas, C.N., and Gerba, C.P. (1991) "Modeling the Risk from *Giardia* and Viruses in Drinking Water," *Journal, AWWA*, **83**, 76-84.

Rein, D.A.; Jamesson, G.M., and Monteith, R.A. (1992) "Toxicity Effects of Alternate Disinfection Processes," In Abstracts, *Water Environment Federation 65th Conference and Exposition,* WEF: Alexandria, VA, 461-471.

Ribo, J.M. and Kaiser, K.L.E. (1987) "Photobacterium Phosphoreum Toxicity Bioassay, I. Test Procedures and Applications," *Toxicological Assessment*, **2**, 305-323.

Rios, R.V.R.A.; da Rocha, L.L.; Vieira, T.G. ; Lago, R.M.; and Augusti, R. (2000) "On-line Monitoring by Membrane Introduction Mass Spectrometry of Chlorination of Organics in Water. Mechanistic and Kinetic Aspects of Chloroform Formation," *Journal of Mass Spectrometry*, **35**, 618-624.

Roefer, P.A.; Monscvitz, J.T.; and Rexing, D.J. (1996) "The Las Vegas Cryptosporidiosis Outbreak," *Journal, AWWA*, **88**, 9, 95.

Rose, J.B.; Dickson, L.J.; Farrah, S.R.; and Carnahan, R.P. (1996) "Removal of Pathogenic and Indicator Microorganisms by a Full-scale Water Reclamation Facility," *Water Research*, **30,** 11, 2785-2797.

Rose, J.B. (1997) "Environmental Ecology of *Cryptosporidium* and Public Health Implications," *Annual Review of Public Health*, **18**, 135-161.

Rose, J.B., Epstein, P.R.; Lipp, E.K.; Sherman, B.H.; Bernard, S.M.; and Patz, J.A. (2001a) "Climate Variability and Change in the United States: Potential Impacts on Water- and Foodborne Diseases Caused by Microbiologic Agents," *Environmental Health Perspectives*, **109**(2), 211-221.

Rose, J.B.; Daeschner, S.; Deasterling, D.R.; Curriero; F.C., Lele; S.; and Patz, J. (2000) "Climate and Waterborne Disease Outbreaks," *Journal, AWWA*, **92**, 77-87.

Rose, J.B. and Grimes, D.J. (2001) *Reevaluation of Microbial Water Quality: Powerful New Tools for Detection and Risk Assessment*, American Academy of Microbiology, Washington, D.C.

Rose, J.B.; Haas, C.N.; and Regli, S. (1991) "Risk Assessment and Control of Waterborne Giardiasis," *American Journal of Public Health*, **81**, 6, 709-713.

Rose, J.B.; Huffman, D.E.; and Gennaccaro, A. (2002) "Risk and control of waterborne cryptosporidiosis," *FEMS Microbiology Reviews*, **26**,113-123.

Rose, J.B.; Huffman, D.E.; Riley, K.; Farrah, S.R.; Lukasik, J.O.; and Harman, C.L. (2001b) "Reduction of Enteric Microorganisms at the Upper Occoquan Sewage authority Water Reclamation Plant," *Water Environment Research*, **73**(6), 711-720

Rosenstein, B.S. (1988) "Repair of Solar UV-induced Damages in Normal Human Skin Fibroblasts," *Photochemistry and Photobiology*, **47**, 275.

Rozen, Y. and Belkin, S. (2001) "Survival of enteric bacteria in seawater," *FEMS Microbiology Reviews*, **25**, 513-529.

Russell, A.; Furr, J.; and Maillard, J. (1997) "Microbial susceptibility and resistance to biocides," *ASM News*, **63**, 481-487.

Saier, M. (1997) "Peter Mitchell and his chemiosmotic theories," *ASM News,* **63**, 13-21.

Saito, T.; Iwase, T.; Horie, J.; and Morioka, T. (1992) "Mode of Photocatalytic Bactericidal Action of Powered Semiconductor Titanium Dioxide on *mutans streptococci,*" *Journal of Photochemistry and Photobiology B*, **14**, 369-379.

Sawyer, C.N. (1957) "Hypochlorination of Sewage," *Sewage and Industrial Wastes*, **29**, 978.

Schmidt, W.; Hambsch, B.; and Petzoldt, H. (1998) "Classification of Algogenic Organic Matter Concerning Its Contribution to the Bacterial Regrowth Potential and By-products Formation," *Water Science & Technology*, **37,** 2, 91-96.

Schurter, L.M.; Bachelor, P.P.; and Margerum, D.W. (1995) "Nonmetal Redox Kinetics: Mono-, Di-, and trichloramine Reactions with Cyanide Ion," *Environmental Science & Technology*, **29**, 4, 1127-1134.

Scully, F.E., Jr. and Hartman, A.C. (1996) "Disinfection Interference in Wastewaters by Natural Organic Nitrogen Compounds," *Environmental Science & Technology*, **30,** 5, 1465-1471.

Scully, F.E., Jr. and Conyers, B. (1997) "Chloramine V: Products and Implications of the Chlorination of Lysine in Municipal Wastewater," *Environmental Science & Technology*, **31**, 1680-1685.

Seidler, R.J.; Morrow, J.E.; and Bagley, S.T. (1977) "*Klebsielleae* in Drinking Water Emanating from Redwood Tanks," *Applied & Envionmental Microbiology*, **33,** 4, 893-900.

Setlow, R.B. (2002) "Shedding Light on Proteins, Nucleic Acids, Cells, Humans and Fish," *Mutation Research*, **511**, 1-14.

Severin, B.F. and Churn, C.C. (1987) "Kinetics of microbial inactivation," AWWA seminar 1-14.

Shang, C. and Blatchley III, E.R. (1999) "Differentiation and Quantification of Free Chlorine and Inorganic Chloramines in Aqueous Solution by MIMS," *Environmental Science & Technology*, **33,** 13, 2218-2223.

Shang, C.; Gong, W-L.; and Blatchley III, E.R. (2000) "Breakpoint Chemistry and Volatile Byproduct Formation Resulting from Chlorination of Model Organic-N Compounds," *Environmental Science & Technology*, **34,** 9, 1721-1728.

Shapiro, H. (1990) "Cell membrane potential analysis," pp. 25-35 in *Methods in Cell Biology*, Academic Press, New York.

Shuler, M.L. and Kargi, F. (1992a) *Bioprocess Engineering: Basic Concepts*, *Prentice-Hall International Editions*, p. 160.

Shuler, M.L. and Kargi, F. (1992b) *Bioprocess Engineering: Basic Concepts*, *Prentice-Hall International Editions*, p. 204.

Silva, A.M.M.; Novelli, E.L.B.; Fascineli, M.L.; and Almeida, J.A. (1999) "Impact of an Environmentally Realistic Intake of Water Contaminants and Superoxide Formation on Tissues of Rats," *Environmental Pollution*, **105,** 2, 243-249.

Sinton, L.W.; Davies-Colley, R.J.; and Bell, R.G. (1994) "Inactivation of enterococci and fecal coliforms from sewage and meatworks effluents in seawater chambers," *Applied & Environmental Microbiology*, **60**, 2040-2048.

Smith, H.V. and Rose, J.B. (1998) "Waterborne *Cryptosporidiosis:* Current Status," *Parasitol Today*, **114**, 14-22.

Sobsey, M.D. (1989) "Inactivation of Health-Related Microorganisms in Water by Disinfection Processes," *Water Science & Technology*, **21**, 3, 179-195.

Sokal, R.R. and Rohlf, F.J. (1981) *Biometry* W.H. Freeman & Co., New York.

Solic, M. and Krstulovic, N. (1992) "Separate and combined effects of solar radiation, temperature, salinity, and pH on the survival of faecal coliforms in seawater," *Marine Pollution Bulletin*, **24**, 411-416.

Standard Methods for the Examination of water and Wastewater, 19th ed. (1995) APHA-AWWA-WEF, Washington, D.C.

Stetler, R.E. (1984) "Coliphages as indicators of enteroviruses," *Applied & Environmental Microbiology*, **41**, 668-670.

Stewart, M.H. and Olson, B.H. (1992) "Impact of growth conditions on resistance of *Klebsiella pneumoniae* to chloramines," *Applied & Environmental Microbiology*, **58**, 2649-2653.

Stewart, P.; Griebe, T.; Srinivasan, R.; Chen, C. I.; Yu, F.; DeBeer, D.; and McFeters, G. (1994) "Comparison of respiratory activity and culturability during monochloramine disinfection of binary population biofilms," *Applied & Environmental Microbiology*, **60**, 1690-1692.

Stott, R. (2002) "Predation of *Cryptosporidium* Oocysts by Protozoa and Rotifers. Implications for Water Quality and Public Health", *Paper presented in IWA 2002 Biennial Conference*, Melbourne, Australia,.

Sung, R.D. (1974) *Effects of Organic Constituents in Wastewater on the Chlorination Process*, Ph. D. Dissertation, Univ. of California, Davis, CA.

Swango, L.J.; Wilt, G.R.; Killen, A.D.; William D.E.; and Worley, S.D. (1987) "Inactivation of *Legionella pneumophila* by Hypochlorite and an Organic Chloramines," *Applied & Environmental Microbiology*, **53,** 12, 2983-2986.

Tanji Y.; Mizoguchi, K.; Yoichi, M.; Morita, M.; Kijima, N.; Kator, H.; and Unno, H. (2003) "Seasonal Change and Fate of Coliphages Infected to *Escherichia coli* 0157:H7 in a Wastewater Treatment Plant," *Water Research*, **37**, 1136-1142.

Tchobanoglous, G. and Burton, F.L. (1991) *Wastewater Engineering: Treatment, Disposal, and Reuse*, McGraw-Hill.

Terzieva, S. and McFeters, G. (1991) "Survival and injury of *Escherichia coli, Campylobacter jejuni*, and *Yersinia enterocolitica* in stream water," *Canadian Journal of Microbiology*, **37**, 785-790.

Terzieva, S. and McFeters, G. (1992) "Effect of chlorine injury on some virulence-related properties of *Campylobacter jejuni," International Journal of Environmental Health Research*, **2**, 24-32.

Torrentera, L., Uribe, R.M.; and Rodriguez, R.R. (1994) "Physical and Biological Characterization of a Seawater Ultraviolet Radiation Sterilizer," *Radiation Physics and Chemistry*, **43**, 3, 249.

Tosa, K.; Hirata, T.; and Taguchi, K. (1995) "Chloramine-induced Injury of Enterotoxigenic *Escherichia coli,*" *Water Science &Technology*, **31,** 5-6, 135-139.

Tracy, H.W.; Camarena, V.M.; and Wing, F. (1966) "Coliform persistence in highly chlorinated water," *Journal, AWWA*, **58**, 9, 1151-1159.

Tree, J.A.; Adams, M.R.; and Lees, D.N. (2003) "Chlorination of Indicator Bacteria and Viruses in Primary Sewage Effluent," *Applied & Environmental Microbiology*, **69**, 4, 2038-2043.

Trehy, M.L., Yost, R.A. and Miles, C.J. (1986) "Chlorination Byproducts of Amino Acids in Natural Waters," *Environmental Science & Technology*, **20**, 1117-1122.

Ueno, H.; Moto, T.; Sayato, Y.; and Nakamuro, K. (1996) "Disinfection By-Products in the Chlorination of Organic Nitrogen Compounds: By-products from Kynurenine," *Chemosphere*, **33**, 1425-1433.

U.S. Environmental Protection Agency (1984) "Ambient water quality criteria for chlorine," *EPA 440/5-84-030,* U.S. Government Printing Office, Washington, D.C.

U.S. Environmental Protection Agency (1997) N-nitrosodimethylamine CASRN 62-75-9, Integrated Risk Information Service (IRIS) Substance File.

U.S. Environmental Protection Agency (1998) *How Wastewater Treatment Works...The Basics.* Office of Water, EPA, Washington, D.C.

U.S. Environmental Protection Agency (2001) *Method 1062: Male-Specific (F+) and Somatic Coliphage in Water by Single Agar Layer (SAL) Procedure*, 821-R-01-029.

U.S. Environmental Protection Agency (2002) *Summary of the August 14-15, 2002, Experts Workshop on Public Health Impacts of Sewer Overflows*, EPA Office of Wastewater Management, Washington, D.C.

Vallentyne, J.R. (1964) "Biogeochemistry of Organic Matter. Part 2, Thermal Reaction Kinetics and Transformation Products of Amino Acids." *Geochimica et Cosmochimica Acta*, 28: 157–88.

Vatsala, S.; Bansal, V.; Tuli, D.K.; Rai, M.M.; Jain, S.K.; Srivastava, S.P.; and Bhatnagar, A.K. (1994) "Gas Chromatographic Determination of Residual Hydrazine and Morpholine in Boiler Feed Water and Steam Condensates," *Chromatographia,* **38**, 7/8, 456-460.

Vives-Rego, J.; Lebaron, P.; and Nebe-von Caron, G. (2000) "Current and future applications of flow cytometry in aquatic microbiology," *FEMS Microbiological Reviews*, **24**, 429-448.

Watanabe, H.; Kitamura, T.; Ochi, S.; and Ozaki, M. (1997) "Inactivation of Pathogenic Bacteria Under Mesophilic and Thermophilic Conditions," *Water Science & Technology*, **36**, 6-7, 25-32.

Weber, D. (1995) "Alternative Disinfection Systems: Regulatory, Design and Permitting Issues," *WEFTEC '95, 68th Annual Conference and Exposition*, **5**, 407.

Weil, I. And Morris, J.C. (1949) "Kinetics Study on the Chloramines: I. The Rates of Formation of Monochloramine, N-chloromethylamine and N-Chlorodimethylamine," *Journal of the American Chemistry Society*, **171**, 1664.

White, G.C. (1986) *The Handbook of Chlorination.* Van Nostrand Reinhold, New York.

William, D.E.; Worley, S.D.; Wheatley, W.B.; and Swango, L.J. (1985) "Bactericidal Properties of a New Water Disinfectant," *Applied & Environmental Microbiology*, **49,** 3, 637-643.

Williamson, K.J. and Nelson, P.O. (1983) "Bacterial Bioassay for Level One Toxicity Assessment," *U.S. Environmental Protection Agency, Enviromental Research Laboratory, Corvallis, OR*. EPA –600/3-83-017.

Wolfe, R.L.; Ward, N.R.; and Olson, B.H. (1985) "Interference in Bactericidal Properties of Inorganic Chloramines by Organic Nitrogen Compounds," *Environmental Science & Technology*, **19,** 12, 1192-1195.

Wolfe, R.S.; Ward, N.R.; Olson, B.H. (1985) "Inactivation of heterotrophic bacterial populations in finished drinking water by chlorine and chloramines," *Water Research*, **19**, 1393-1403.

Wong, P.S.H.; Cooks, R.G.; Cisper, M. E.; and Hemberger, P.H. (1995) "On-line *In Situ* Analysis with Membrane Introduction MS," *Environmental Science & Technology*, **29**, 215A-218A.

World Health Organization (1993) *Guidelines for Drinking Water Quality*, 2nd ed., World Health Organization; Geneva, **1**.

Yang, L.; Chang, W-S.; and Huang, M-N (2000) "Natural Disinfection of Wastewater in Marine Outfall Fields," *Water Research*, **34**, 3, 743-750.

Yasunori, T.; Mizoguchi, K.; Yoichi, M.; Morita, M.; Kijima, N.; Kator, H.; and Unno, H. (2003) "Seasonal Change and Fate of Coliphages Infected to *Escherichia coli* 0157:H7 in a Wastewater Treatment Plant," *Water Research*, **37**, 1136-1142.

Yeh, H.H.; Tseng, I.C.; and Lai, W.L. (1998) "Chlorine Residual and Assimilable Organic Carbon (AOC) for Drinking Water Quality Control in Taiwan," *Water Supply*, **16**, 237-243.

Yoon, J. and Jensen, J.N. (1995) "Chlorine Transfer From Inorganic Monochloramine in Chlorinated Wastewater," *Water Environment Research*, **67**, 5, 842-847.

Yu, F.P. and McFeters, G. (1994) "Physiological responses of bacteria in biofilms to disinfection," *Applied & Environmental Microbiology*, **60**, 2462-2466.

Zaske, S.K.; Dockins, W.S.; Schillinger, J.E.; and McFeters, G.A. (1980) "New methods to assess bacterial injury in water," *Applied & Environmental Microbiology*, **39**, 656-658.

Zobell, C.E. (1943) "The effect of solid surfaces upon bacterial activity," *Journal of Bacteriology*, **46**, 39-56.

Zwart, D. and Slooff, W. (1983) "The Microtox as an Alternative Assay in the Acute Toxicity Assessment of Water Pollutants," *Aquatic Toxicology*, **4**, 129-138.

WASTEWATER UTILITY

Alabama
Montgomery Water Works & Sanitary Sewer Board

Alaska
Anchorage Water & Wastewater Utility

Arizona
Gila Resources
Glendale, City of, Utilities Department
Mesa, City of
Peoria, City of
Pima County Wastewater Management

Arkansas
Little Rock Wastewater Utility

California
Calaveras County Water District
Central Contra Costa Sanitary District
Crestline Sanitation District
Delta Diablo Sanitation District
Dublin San Ramon Services District
East Bay Dischargers Authority
East Bay Municipal Utility District
Eastern Municipal Water District
El Dorado Irrigation District
Fairfield-Suisun Sewer District
Fort Lauderdale, City of
Irvine Ranch Water District
Las Virgenes Municipal Water District
Livermore, City of
Lodi, City of
Los Angeles, City of
Los Angeles County, Sanitation Districts of
Napa Sanitation District
Orange County Sanitation District
Palo Alto, City of
Riverside, City of
Sacramento Regional County Sanitation District
San Diego Metropolitan Wastewater Department, City of
San Francisco, City & County of
San Jose, City of
Santa Barbara, City of
Santa Cruz, City of
Santa Rosa, City of
South Bayside System Authority
South Coast Water District
South Orange County Wastewater Authority
Sunnyvale, City of
Union Sanitary District
West Valley Sanitation District

Colorado
Boulder, City of
Colorado Springs Utilities
Greeley, City of
Littleton/Englewood Water Pollution Control Plant
Metro Wastewater Reclamation District, Denver

Connecticut
The Mattabassett District
New Haven, City of, WPCA

District of Columbia
District of Columbia Water & Sewer Authority

Florida
Broward, County of
Fort Lauderdale, City of
Gainesville Regional Utilities
JEA
Miami-Dade Water & Sewer Authority
Orange County Utilities Department
Reedy Creek Improvement District
Seminole County Environmental Services
St. Petersburg, City of
Stuart Public Utilities
Tallahassee, City of
Tampa, City of
Toho Water Authority
West Palm Beach, City of

Georgia
Atlanta Department of Watershed Management
Augusta, City of
Clayton County Water Authority
Cobb County Water System
Columbus Water Works
Fulton County
Gwinnett County Department of Public Utilities
Savannah, City of

Hawaii
Honolulu, City & County of

Idaho
Boise, City of

Illinois
American Bottoms Wastewater Treatment Plant
Greater Peoria Sanitary District
Kankakee River Metropolitan Agency
Metropolitan Water Reclamation District of Greater Chicago
Wheaton Sanitary District

Iowa
Ames, City of
Des Moines Metro Wastewater Reclamation Authority
Iowa City

Kansas
Johnson County Unified Wastewater Districts
Unified Government of Wyandotte County/ Kansas City, City of

Kentucky
Louisville & Jefferson County Metropolitan Sewer District

Louisiana
Sewerage & Water Board of New Orleans

Maine
Bangor, City of
Portland Water District

Maryland
Anne Arundel County Bureau of Utility Operations
Howard County Department of Public Works
Washington Suburban Sanitary Commission

Massachusetts
Boston Water & Sewer Commission
Upper Blackstone Water Pollution Abatement District

Michigan
Ann Arbor, City of
Detroit, City of
Holland Board of Public Works
Lansing, City of
Saginaw, City of
Wayne County Department of Environment
Wyoming, City of

Minnesota
Rochester, City of
Western Lake Superior Sanitary District

Missouri
Independence, City of
Kansas City Missouri Water Services Department
Little Blue Valley Sewer District
Metropolitan St. Louis Sewer District

Nebraska
Lincoln Wastewater System

Nevada
Henderson, City of

New Jersey
Bergen County Utilities Authority
Passaic Valley Sewerage Commissioners

New York
New York City Department of Environmental Protection

North Carolina
Charlotte/Mecklenburg Utilities
Durham, City of
Metropolitan Sewerage District of Buncombe County
Orange Water & Sewer Authority

Ohio
Akron, City of
Butler County Department of Environmental Services
Columbus, City of
Metropolitan Sewer District of Greater Cincinnati
Northeast Ohio Regional Sewer District
Summit, County of

Oklahoma
Trinity River Authority
Tulsa, City of

Oregon
Clean Water Services
Eugene/Springfield Water Pollution Control
Water Environment Services

Pennsylvania
Philadelphia, City of
University Area Joint Authority

South Carolina
Charleston Commissioners of Public Works
Mount Pleasant Waterworks & Sewer Commission
Spartanburg Sanitary Sewer District

Tennessee
Cleveland, City of
Knoxville Utilities Board
Murfreesboro Water & Sewer Department
Nashville Metro Water Services

Texas
Austin, City of
Dallas Water Utilities
Denton, City of
El Paso Water Utilities
Fort Worth, City of
Gulf Coast Waste Disposal Authority
Houston, City of
San Antonio Water System
Trinity River Authority

Utah
Salt Lake City Corporation

Virginia
Alexandria Sanitation Authority
Arlington, County of
Fairfax County
Hampton Roads Sanitation District
Henrico, County of

Hopewell Regional Wastewater Treatment Facility
Loudoun County Sanitation Authority
Lynchburg Regional WWTP
Prince William County Service Authority
Richmond, City of
Rivanna Water & Sewer Authority

Washington
Bellevue Utilities Department
Edmonds, City of
Everett, City of
King County Department of Natural Resources
Seattle Public Utilities
Sunnyside, Port of
Yakima, City of

Wisconsin
Green Bay Metro Sewerage District
Madison Metropolitan Sewerage District
Milwaukee Metropolitan Sewerage District
Racine, City of
Sheboygan Regional Wastewater Treatment
Wausau Water Works

Australia
South Australian Water Corporation
Sydney Water Corporation
Water Corporation of Western Australia

Canada
Greater Vancouver Regional District
Regina, City of, Saskatchewan
Toronto, City of, Ontario
Winnipeg, City of, Manitoba

Mexico
Servicios de Agua y Drenaje de Monterrey, I.P.D.

New Zealand
Watercare Services Limited

United Kingdom
Yorkshire Water Services Limited

STORMWATER UTILITY

California
Los Angeles, City of, Department of Public Works
Monterey, City of
San Francisco, City & County of
Santa Rosa, City of

Colorado
Boulder, City of

Georgia
Griffin, City of

Iowa
Cedar Rapids Wastewater Facility
Des Moines Metro Wastewater Reclamation Authority

Kansas
Overland Park, City of

Maine
Portland Water District

North Carolina
Charlotte, City of, Stormwater Services

Pennsylvania
Philadelphia, City of

Tennessee
Chattanooga Stormwater Management

Washington
Seattle Public Utilities

STATE

Arkansas Department of Environmental Quality
Fresno Department of Public Utilities
Kansas Department of Health & Environment
Urban Drainage & Flood Control District, CO

CORPORATE

ADS Environmental Services
The ADVENT Group Inc.
Alan Plummer & Associates
Alden Research Laboratory Inc.
Aqua-Aerobic Systems Inc.
Aquateam–Norwegian Water Technology Centre A/S
ARCADIS
Associated Engineering
BaySaver Inc.
Black & Veatch
Blasland, Bouck & Lee Inc.
Boyle Engineering Corporation
Brown & Caldwell
Burns & McDonnell
CABE Associates Inc.
The Cadmus Group
Camp Dresser & McKee Inc.
Carollo Engineers Inc.
Carpenter Environmental Associates Inc.
CDS Technologies Inc.
Chemtrac Systems Inc.
CH2M HILL
Damon S. Williams Associates, LLC
David L. Sheridan, P.C.
Dewling Associates Inc.
Earth Tech Inc.
Ecovation
EMA Inc.
The Eshelman Company Inc.
EWT Holdings Corporation
Fay, Spofford, & Thorndike Inc.
Freese & Nichols Inc.
ftn Associates Inc.
Fuss & O'Neill Inc.
Gannett Fleming Inc.
GE Betz
Greeley and Hansen LLC
Hazen & Sawyer, P.C.
HDR Engineering Inc.
HNTB Corporation
HydroQual Inc.
Infilco Degremont Inc.
Ingersoll-Rand Energy Systems
Insituform Technologies Inc.
Institute for Environmental Technology & Industry, Korea
Jacobson Helgoth Consultants Inc.
Jason Consultants LLC Inc.
Jordan, Jones, & Goulding Inc.
KCI Technologies Inc.
Kelly & Weaver, P.C.
Kennedy/Jenks Consultants
Komline Sanderson Engineering Corporation
Lawler, Matusky & Skelly Engineers, LLP
LTI Inc.
Lombardo Associates Inc.
Malcolm Pirnie Inc.
McKim & Creed
MEC Analytical Systems Inc.
Metcalf & Eddy Inc.
MWH
New England Organics
O'Brien & Gere Engineers Inc.
Odor & Corrosion Technology Consultants Inc.
Oswald Green, LLC
PA Government Services Inc.
Parametrix Inc.
Parsons
Post, Buckley, Schuh & Jernigan
R&D Engineering/Conestoga Rover & Associates
RMC Inc.
R.M. Towill Corporation
Ross & Associates Ltd.
Royce Technologies
SAIC Maritime Technical Group
Short Elliott Hendrickson Inc.
Stantec Consulting Inc.
Stearns & Wheler, LLC
Stone Environmental Inc.
Stormwater Management Inc.
Stratus Consulting Inc.
Synagro Technologies Inc.
Tetra Tech Inc.
Trojan Technologies Inc.
URS Corporation
USfilter
Veola Water North America
Wade-Trim Inc.
Westin Engineering Inc.
Weston Solutions Inc.
Woodard & Curran
WRc/D&B, LLC
WWETCO, LLC
Zenon Environmental Inc.
Zoeller Pump Company

INDUSTRY

American Electric Power
ChevronTexaco Energy Research & Technology Company
The Coca-Cola Company
Dow Chemical Company
DuPont Company
Eastman Chemical company
Eastman Kodak Company
Eli Lilly & Company
Merck & Company Inc.
ONDEO Services
Procter & Gamble Company
PSEG Services Corporation
RWE Thames Water Plc
Severn Trent Services Inc.
United Water Services LLC

Note: List as of 2/1/05

IWA Publishing's authorised EU representative for General Product Safety Regulations is Diane D'Arras, 15 rue Duret, 75116 Paris, France, e-mail: safety@iwap.co.uk.

Printed and bound by CPI Group (UK) Ltd, Croydon, CR0 4YY
12/05/2026
02108868-0006